$h = 6.626 \times 10^{-34}$ Js (Planck)

$\hbar = \frac{h}{2\pi} = 1.05 \times 10^{-34}$

$k_B = 1.381 \times 10^{-23}$ JK^{-1} (Boltzmann)

$r_B = 5.3 \times 10^{-11}$ m (0.53 Å)

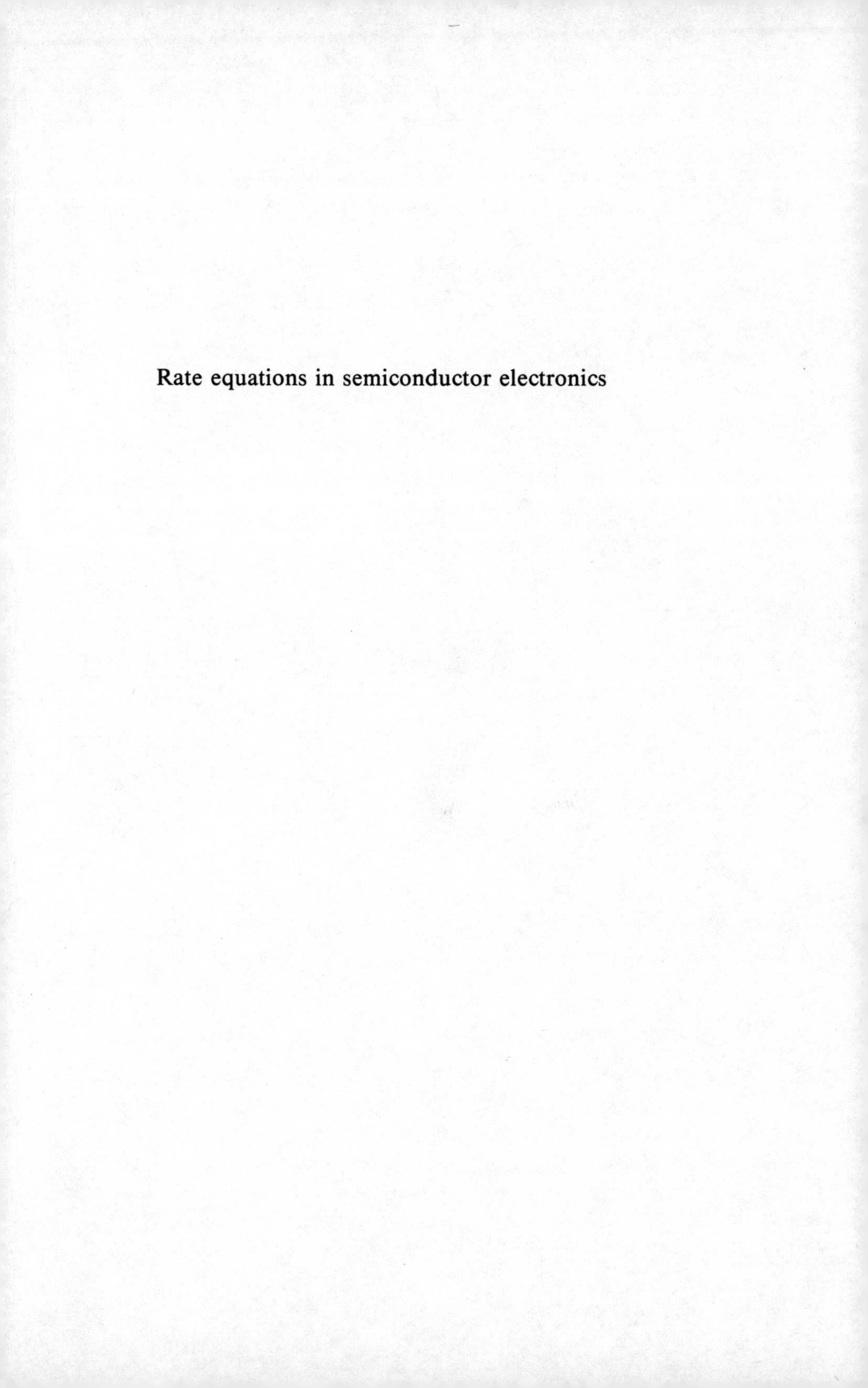

Rate equations in semiconductor electronics

Rate equations in semiconductor electronics

J. E. CARROLL

Professor of Engineering, University of Cambridge and Fellow of Queen's College

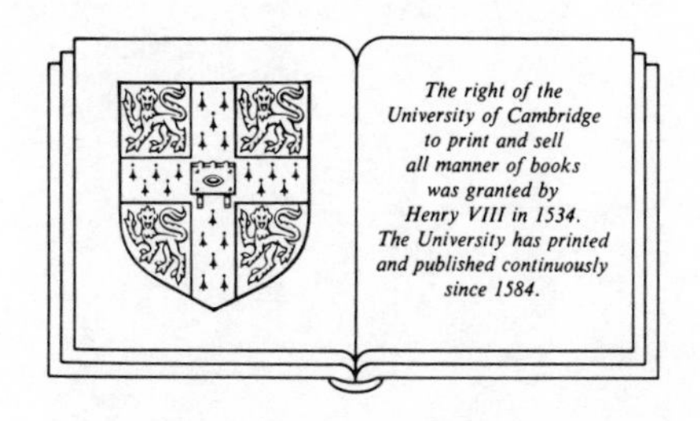

CAMBRIDGE UNIVERSITY PRESS

Cambridge

London New York New Rochelle

Melbourne Sydney

Published by the Press Syndicate of the University of Cambridge
The Pitt Building, Trumpington Street, Cambridge CB2 1RP
32 East 57th Street, New York, NY 10022, USA
10 Stamford Road, Oakleigh, Melbourne 3166, Australia

First published 1985

Printed in Great Britian by
J. W. Arrowsmith Ltd., Bristol

British Library cataloguing in publication data

Carroll, John E. (John Edward) 1934–
Rate equations in semiconductor electronics.

1. Semiconductors - Mathematics
I. Title
537.6′22 QC611.24

Library of Congress cataloguing in publication data

Carrroll, John E. (John Edward), 1934–
Rate equations in semiconductor electronics.

Bibliography: p.
Includes index.
1. Semiconductors - Mathematics. 2. Optoelectronics - Mathematics. I. Title.
TK7871.85.C284 1985 621.381′042 85-5887

ISBN 0 521 26533 9

AS

TO VERA
for 25 years at a rapid rate

CONTENTS

PREFACE

Collecting fresh fruits becomes ever harder as the tree of knowledge grows higher and wider. However, there are certain branches that provide surer footholds to the new growths, and teachers must search these out. The rates of change of charge, energy, momentum, photon numbers, electron densities and so on, along with their detailed balances as particles and systems interact, provide fundamental footholds on sturdy branches in physics and chemistry. This book contains a collection of such topics applied to semiconductors and optoelectronics in the belief that such analyses provide valuable tutorial routes to understanding past, present and future devices. By concentrating on rates of change one focuses attention on these devices' dynamic behaviour which is vital to the ever faster flow of digits and information. The rates of the statistical interactions between electrons and photons determine distributions of energy amongst the particles as well as determining distributions in time, so controlling the ways in which devices work.

The first chapter is meant to be a fun chapter outlining some of the breadth and ideas of rate equation approaches. It is even hoped that some of these initial ideas may be picked up by sixth form teachers. Rates of reactions are mentioned in school chemistry but the implications are much broader.

Most electronic degree courses consider electron waves, holes and electrons, along with devices such as p–n junctions, FETS and bipolar transistors. Chapters 2 and 3 are adjuncts to this work. By considering rates of change of charge and emphasising transit times and recombination rates the dynamics of these devices can be highlighted.

Chapter 4 contains in its initial parts the basis for the charge transport equation in a semiconductor, work which is covered in the third year of Electrical Sciences at Cambridge. The electron transfer effect discovered experimentally by J. B. Gunn, who also studied at Cambridge, is considered in this chapter, at a level for elective courses. Chapters 5 and 6 contain material on energy distributions for electrons and photons and how they interact. This material is again covered in final year teaching. It is hoped that the unified rate equation approach to lasers, LEDs and photodiodes in spite of its simplifications has useful tutorial value in many optoelectronic courses.

Chapter 7 contains topics for elective final year degree courses and first year research lectures. The inclusion of this material seems essential to demonstrate the exciting extent to which rate equations can take a discussion, along with some of their limitations. The development of rate equations from Maxwell's equations should be especially helpful in forming links between quantum and classical discussions.

The author then has three main hopes for this book: (i) That university and polytechnic lecturers will buy and recommend this book as an adjunct to a range of existing degree courses, in semiconductors and in optoelectronics. (ii) That degree course students studying semiconductors and optoelectronics will find the book helpful alongside their normal texts. Problems and solution notes are provided to help private study. (iii) That research and development departments of the optoelectronic industry will buy the book and find that rate equation concepts will help invention and the formation of technical judgments for future devices where the faster flow of information in communication systems puts stress on the dynamic performance.

As ever, I am grateful to all my colleagues who have taught me so much. I am pleased to acknowledge help from Herman Haus who corresponded with me over rate equations for mode locking. I am delighted to single out Alan Beardsworth who as a project student helped programme a BBC microcomputer for the electron transfer effect. Indeed I am grateful to all my students and to a system where I have had to supervise undergraduates in small groups with their many problems. I do not know how much I have taught them, but with their close questioning I felt at the end that I had understood it all!

I would like to thank Irene Pizzie of Cambridge University Press for a very close reading of the manuscript. The errors that will remain will be my fault not hers.

Cambridge 1985

1

Introduction to rate equations

1.1 Introduction

An outstandingly innovative scientist, Rudi Kompfner, wrote that when his intuition was unengaged or disengaged then his creative faculties were paralysed.[1] Although Kompfner was writing about quantum theory, his remarks apply to most aspects of science. How can one create and innovate when no understanding is present? The idea behind this book is that a useful contribution to understanding in science and engineering can be found by determining the rate at which an interactive process occurs and concentrating on the dominant features which limit the interaction rates.

Such thinking is not limited to science; it can have universal application. For example, before lending money to a client, a building society will ask how much that client is earning from any employers, and so obtain an estimate of the maximum rate at which the client can reasonably pay off the mortgage that will be advanced to buy a house. The rate of income being paid to the client determines to a first order the rate at which money can be spent! The maximum amount of traffic that can use a road may be limited not by the size of the road, but by the rate at which traffic can escape or enter from congested roundabouts that serve the road. In building electronic circuits to switch at high speeds, one may find that the speed is limited by the rate at which components can transfer charge into a capacitive load. It may alternatively be limited by the rate at which information can be transmitted from neighbouring devices, which have to be a certain distance away in order to accommodate enough devices to drive and be driven by any one single device.

The ubiquity of rate equations means that it is inevitable that a selection has to be made of the topics to be covered. The selection here is naturally idiosyncratic to the author's experience in semiconductor and optical devices, where rate equation concepts seem to be particularly useful in understanding some of the dynamics of devices and the interactions of photons and electrons. The statistical implications of rate equations in energy as well as time are considered. The emphasis is on tutorial models rather than completely rigorous models. It is hoped that the text will help to shed light on the way things work, show how light may be generated from work (as one example) and make light of the work of many problems. This initial chapter takes a light hearted look at many of the areas to which rate equation thinking can be applied.

1.2 Continuity

Every driver knows that if the flow of traffic into a road exceeds the rate at which the cars flow out then a traffic jam can occur. Conservation of cars, conservation of mass, conservation of energy, and so on lead to one of the most fundamental rate equations: continuity of matter.

Suppose that on the eastbound lane of a motorway at x, there is a junction where, on average there are $J(x)$ cars joining every second (Fig. 1.1). At the next junction at $x+d$ there are $J(x+d)$ cars per second leaving the lane. It follows that for the C cars on the lane between the two junctions:

$$\partial C/\partial t = J(x) - J(x+d) \tag{1.2.1}$$

Define an average density of cars from $\rho = C/d$, with d being the distance between the two junctions. Expand $J(x+d)$ to a *first* order in d:

$$\begin{aligned} d\ \partial\rho/\partial t &= J(x) - [J(x) + d\ \partial J/\partial x] \\ &= -d\ \partial J/\partial x \end{aligned}$$

or

$$(\partial\rho/\partial t) + (\partial J/\partial x) = 0 \tag{1.2.2}$$

This is the classic one-dimensional equation of continuity relating the flux of particles (here cars) with their increase in density. Matter which is not created or destroyed must, having entered a region, either flow out again or accumulate. The continuity equation gives the rate of accumulation.

Fig. 1.1. Continuity in one dimension. $J(x)$ cars joining eastbound route at x and $J(x+d)$ cars leaving at $x+d$.

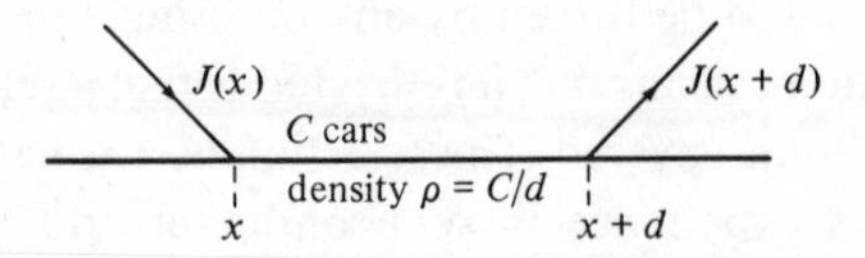

The same equation holds for the flow of charged particles such as electrons with n particles per unit volume. Here an electrical current density J_x replaces J in (1.2.2) as the current flow in the direction x with

$$J_x = \rho v_x \quad \text{where } \rho = en \tag{1.2.3}$$

n being the particle density but ρ being the charge density. For electrons we adopt the notation $e = -q$, where q is the magnitude of the electronic charge (1.6×10^{-19}C).

In more than one dimension, the standard technique considers a cube of volume $\mathrm{d}V$ (Fig. 1.2) centred at x, y, z in which there is a charge $Q = \rho\,\mathrm{d}V = \rho\,\mathrm{d}x\,\mathrm{d}y\,\mathrm{d}z$. Current flowing in at the middle of the face is $[J_x(x - \frac{1}{2}\mathrm{d}x, y, z)\,\mathrm{d}y\,\mathrm{d}z]$ with a current $[J_x(x + \frac{1}{2}\mathrm{d}x, y, z)\,\mathrm{d}y\,\mathrm{d}z]$ flowing out, where $\mathrm{d}y\,\mathrm{d}z$ is the area at both $x - \frac{1}{2}\mathrm{d}x$ and $x + \frac{1}{2}\mathrm{d}x$. Repeating the calculation in a similar way for the current flow at the centre of each facet of the elemental volume gives

$$\begin{aligned}\partial Q/\partial t = {} & [J_x(x - \tfrac{1}{2}\mathrm{d}x, y, z)\,\mathrm{d}y\,\mathrm{d}z] - [J_x(x + \tfrac{1}{2}\mathrm{d}x, y, z)\,\mathrm{d}y\,\mathrm{d}z] \\ & + [J_y(x, y - \tfrac{1}{2}\mathrm{d}y, z)\,\mathrm{d}z\,\mathrm{d}x] - [J_y(x, y + \tfrac{1}{2}\mathrm{d}y, z)\,\mathrm{d}z\,\mathrm{d}x] \\ & + [J_x(x, y, z - \tfrac{1}{2}\mathrm{d}z)\,\mathrm{d}x\,\mathrm{d}y] - [J_x(x, y, z + \tfrac{1}{2}\mathrm{d}z)\,\mathrm{d}x\,\mathrm{d}y]\end{aligned}$$

(and expanding to the first order in $\mathrm{d}x\,\mathrm{d}y\,\mathrm{d}z$)

$$\partial Q/\partial t = -[(\partial J_x/\partial x) + (\partial J_y/\partial y) + (\partial J_z/\partial z)]\,\mathrm{d}x\,\mathrm{d}y\,\mathrm{d}z \tag{1.2.4}$$

Cancelling the elemental volume $\mathrm{d}V = \mathrm{d}x\,\mathrm{d}y\,\mathrm{d}z$ from both sides:

$$\partial\rho/\partial t = -\mathrm{div}\,\boldsymbol{J} \tag{1.2.5}$$

Fig. 1.2. Continuity in three dimensions (notation). Elemental cube of sides $\mathrm{d}x\,\mathrm{d}y\,\mathrm{d}z$ centred at x, y, z, filled with density ρ. J_x is the x-component of the particle current $\boldsymbol{J}$; similarly for the other two dimensions.

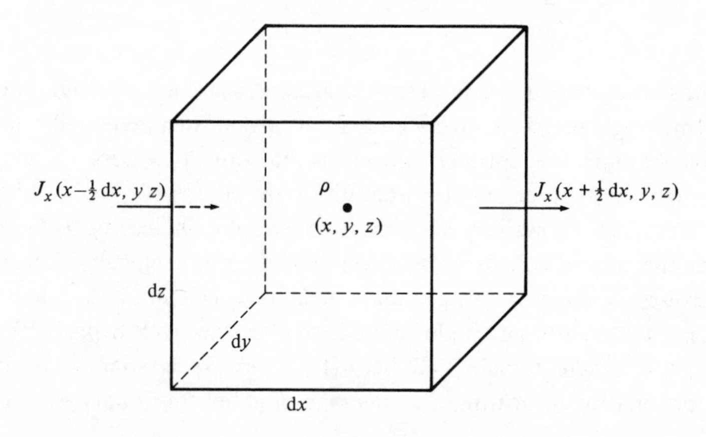

where

$$\text{div}\,\boldsymbol{J} = [(\partial J_x/\partial x) + (\partial J_y/\partial y) + (\partial J_z/\partial z)]$$

For any flow where the divergence of the current flow is zero, there can be no accumulation of charge. Similarly, where mass is conserved there can be no accumulation of mass. Such continuity relations[6] will be used time and again.

The continuity of energy flow also has to be considered in the same way because energy is not created or destroyed. Power $\boldsymbol{P}$ is the rate of flow of energy per unit area and takes the role of energy current density. The energy per unit volume is the energy density U. The continuity relation then has to be

$$\partial U/\partial t = -\text{div}\,\boldsymbol{P} \tag{1.2.6}$$

where $\boldsymbol{P} = U\boldsymbol{v}_g$ with $\boldsymbol{v}_g$ being the velocity at which the energy travels. For particles of density n, $U = n\mathscr{E}$, where $\mathscr{E}$ is the main energy associated with the particles. A similar result is found for electromagnetic theory (section 7.5).

1.3 Chemical reactions

Most readers will have studied elementary chemistry[2] where one is told that 'the rate of reaction is proportional to the concentration of the components taking part in that reaction'. This basic tenet presupposes a certain classical form of interaction which will be met time and time again in this book. It may be helpful to start to think about this law in a simple way to make it easier to understand its likely limitations.

A light hearted examination of the law may be given by considering the rate at which couples get married. In 'chemical' terms one might write the result of an unattached male (m) plus an unattached female (f) combining to give a married couple (M) as:

$$m + f \rightarrow M \tag{1.3.1}$$

Suppose then that in a country, of unit area, there are [f] single females and [m] single males. At every encounter of one such male and one such female (a binary encounter – meaning that only two persons are concerned) there is some small probability p that the encounter will lead to marriage (the formation of a married couple, increasing [M] by 1). Assuming a well mixed population that does not cluster into sets or subgroups, then every single male will have [f] times the chance of meeting with any one single female, so that the overall probability of meeting a single female will be σ[f], where σ is some constant of proportionality indicating the average number of binary encounters between any one unattached male and an unattached female during a year. However, there are [m] males so that the total number of binary encounters must be $\{[m]\sigma[f]\}$ per year leading to the following average rate of increase of married couples:

$$d[M]/dt = p\sigma[m][f] \tag{1.3.2}$$

with the time scale measured here in years.

The rate of reaction then is proportional to the concentration of the components ([m] single males/unit area and [f] single females/unit area) taking part. Dissociation or break up of married couples would be a further complication which we have not considered.

Consider the chemical reaction between iodine and hydrogen molecules:

$$H_2 + I_2 \rightarrow 2HI \tag{1.3.3}$$

Under certain conditions this reaction is adequately expressed by the rate equation

$$d[HI]/dt = K_+[H_2][I_2] \tag{1.3.4}$$

where K_+ is the constant of proportionality and the square brackets denote the concentration of the chemical. Here the reaction proceeds between random encounters of hydrogen and iodine in a well mixed set of gases. The discussion leading to (1.3.2) then holds. However, if one considers the reaction between hydrogen and bromine, a more complicated rate equation must be expected because the bromine molecule first has to dissociate:

$$Br_2 \rightarrow Br + Br \quad \text{with} \quad Br + H_2 \rightarrow H + HBr \tag{1.3.5}$$

and so on. The form of (1.3.4) will not apply to this reaction.

In general one must allow for reactions to go in either direction. So with hydrogen iodide one should have

$$\text{and} \quad \left.\begin{aligned} H_2 + I_2 &\rightarrow 2HI \\ 2HI &\rightarrow H_2 + I_2 \end{aligned}\right\} \tag{1.3.6}$$

The dissociation of HI into H_2 and I_2 requires two molecules of HI to come together, so one expects this reaction to occur at a rate proportional to $[HI][HI] = [HI]^2$. A more general form for (1.3.4) then should read

$$d[HI]/dt = K_+[H_2][I_2] - K_-[HI]^2 \tag{1.3.7}$$

The much lower magnitude of K_- compared to K_+ at temperatures below 500 K permits (1.3.4) to be used. At temperatures around 800 K, both forward and reverse reactions need to be considered. Problem 1.2 indicates how one can deal with this equation, and in particular how equilibrium conditions can be used to determine the relations between the rate constants.

Simple theories for rate equations rarely give values for the rate constants, such as K_+ or K_-. This is not as serious an indictment as it seems at first sight. Even more detailed theories usually have to rely, at some juncture, on experimental data that may be hard to determine. Rate equations, by giving the form for the equations governing the rates of reaction, can indicate sensible experiments that need to be undertaken in order to evaluate the appropriate constants. Sometimes the relative rates of reaction may be inferred from other data. For example, chlorine is known to be a more reactive substance than iodine so that the reaction

$$H_2 + Cl_2 \rightarrow 2HCl \tag{1.3.8}$$

can be an explosive reaction given enough energy to initiate it. (For example, light from a burning magnesium strip can give enough energy to initiate this reaction.) The reaction of hydrogen and iodine (1.3.3) is slower, in keeping with the lower reactivity of iodine.

More can be said about the constants of proportionality in such rate equations if one thinks about the mechanisms of the reactions as well as the rates. To initiate certain reactions it is often necessary to have an input U_a of energy - the activation energy. At any temperature T, one expects (as we shall see later in the book) that the number of molecules with an energy above U_a will be proportional to $\exp(-U_a/kT)$. As the temperature is increased, so the random kinetic energy amongst the molecules increases and more molecules have enough energy to start the

Fig. 1.3. Arrhenius plots. Logarithm of rate of reaction plotted against reciprocal temperature. Ln $K = -U_a/kT$: the slope determines the activation energy U_a. Data appropriate to rate of reaction of hydrogen iodide [2].

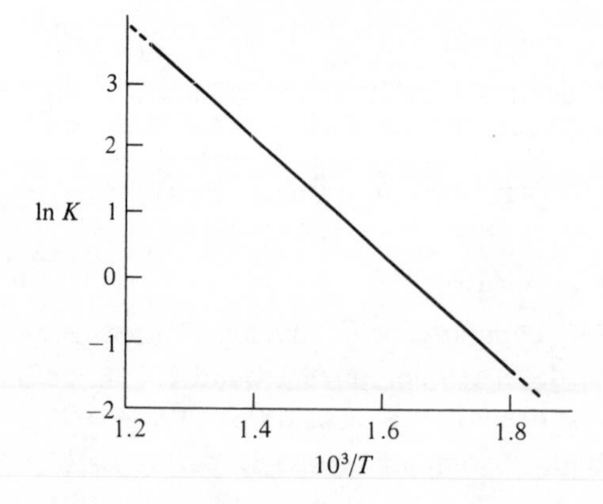

reaction. The rate constant K_+ in (1.3.4) may then be expected to vary exponentially with temperature.

A plot of ln K_+ against $1/T$ in equations like (1.3.4) will often yield an almost linear relationship (see Fig. 1.3), with the slope giving the activation energy of the process (about 41 kcal/mole in the case of (1.3.4)). Such a plot is known as an Arrhenius plot after the Swedish scientist Svante Arrhenius who showed, around 1889, the wide applicability of this relationship.

It is also found that temperature alters the pre-exponential factors through affecting the rate of change of encounters, so that typically a more general result is

$$K \sim AT^n \exp(-U_a/kT) \tag{1.3.9}$$

where n is usually $\frac{1}{2}$.

1.4 Buying and selling houses: quantised transactions

One of the least tangible theories of our time is quantum theory.[7] In mechanics one can feel a force, experience acceleration, observe distances. In electricity, meters can register the flow of current, lamps can light, electrons can be observed to hit scintillating screens. Quantum theory is more abstract[3] and often refers to situations at the extreme ends of experience: dimensions which cannot be observed with the naked eye, exceedingly low light levels, single electrons, and so on. To add to our difficulties the full theory contains many abstract concepts such as vectors in a space of infinite numbers of dimensions.[4]

Rate equations help to bring some understanding into this difficult area by building on a readily grasped concept that energy comes in packets or quanta. The concept of discrete transactions is in fact commonplace, and it may have tutorial value to consider in a light hearted way one such set of discrete transactions - the buying and selling of houses.

The discussion is started with the assumption that each house has only one owner (A. N. Electron) who may either occupy or sell his house. Naturally if he sells it with 'vacant possession' potential buyers regard it as a 'hole'! Consider then the buying and selling of houses that occurs between two areas only: an expensive West-end area of London and some cheaper East-end residential area. In each respective area there are N_w and N_e houses with n_w and n_e occupiers. It follows that there are $[N_w - n_w]$ and $[N_e - n_e]$ vacant sites in the respective areas.

Some spontaneous agreements to buy and sell will be made (Fig. 1.4(a)), and it is observed that these spontaneous exchanges are initiated only from the ranks of the rich West-end occupants (n_w) who have the financial support to purchase directly one of the $[N_e - n_e]$ vacant sites in the East end. The rate at which such spontaneous exchanges can be made is expected, to a first order, to be proportional to the available number of potential buyers $[n_w]$. Because each buyer has the choice of $[N_e - n_e]$ empty sites, spontaneous sales on this simple model would be expected to be proportional to the product $n_w[N_e - n_e]$.

The very fact that spontaneous sales occur means that money is generated, which encourages the firm of estate agents (known as Photon and Photon) with P partners to further stimulate sales from the rich and to generate more money. The greater the number of profitable sales then the greater the number of partners taken on to join the firm. Consequently, the very fact that spontaneous sales occur starts an increase in the number of partners P proportional to these spontaneous sales, so that we would expect:

$$\mathrm{d}[P]/\mathrm{d}T)_{\text{spontaneous}} = A[n_w][N_e - n_e] \tag{1.4.1}$$

for some constant A.

The estate agents themselves vigorously stimulate sales in proportion to the number of partners P in the firm, to the number of potential buyers, and to the number of vacant properties. So for profitable sales to the rich West-enders of the vacant East-end properties there is another

Fig. 1.4. Schematic diagram of house sales (between expensive and less expensive areas). (*a*) Spontaneous sales where finance is not required. (*b*) Stimulated sales which generate money for estate agents *P*. (*c*) Stimulated sales requiring finance from estate agents *P* to enable movement to occur. Sales are proportional to the product of number of buyers and the number of sellers, and also for stimulated sales to numbers of partners in the estate agents. Sellers are shown as leaving vacant possession.

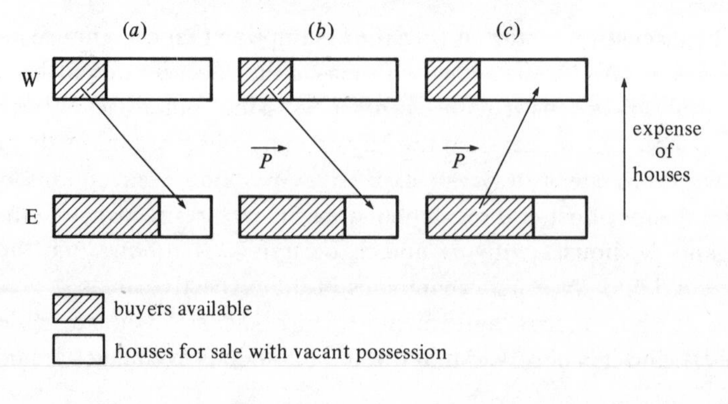

rate of growth such that

$$d[P]/dt)_{\text{stimulated}} = B[P][n_w][N_e - n_e] \tag{1.4.2}$$

for some constant B which is, in general, different from A.

The high value market of $[N_w - n_w]$ vacant West-end sites can be bought by some of the $[n_e]$ possible purchasers in the East end only if they are helped to gain the finance through any one of the P partners who arrange terms and mortgages for property in these two areas. This less profitable business would lead to a reduction in the number of partners if it were the only business to be done, and following much the same arguments as for (1.4.2), but in reverse,

$$d[P]/dt)_{\text{decay}} = -C[P][n_e][N_w - n_w] \tag{1.4.3}$$

Forgetting about losses of partners to other employers and adding all the rates together

$$d[P]/dt = C[P][n_e][n_w]\{[(B/C)+(A/CP)]F_e - F_w\} \tag{1.4.4}$$

where

$$F_{e/w} = [N_{e/w}/n_{e/w}] - 1 \tag{1.4.5}$$

By definition, the equilibrium values give $dP/dt = 0$:

$$P_{equ} = AF_e/[CF_w - BF_e] \tag{1.4.6}$$

If the partners are dealing with two very similar districts then $F_e \sim F_w$ and $P_{equ} \sim A/(C-B)$ (see problem 1.3), with fewer partners for districts where $F_w/F_e \gg 1$. The experimentally observed equilibrium values can help to evaluate the constants A, B, C in well-founded rate equations.

It is hoped that the reader may see some analogies between this interaction and the interactions between electrons changing energy levels stimulated by photons.[5] In chapters 5 and 6 such interactions will be studied in more depth, and it will be seen how the numbers P can fluctuate or even collapse if there are not enough stimulated interactions through lack of electrons or holes (buyers or sellers). Like any analogy, beware of pushing it too far!

1.5 Rate of change of probability

Many practical problems are concerned with random rates of arrival. Letters arrive randomly through the post. Random cars pass by the drive outside the house to prevent a clear exit! In telecommunications, signals arrive with random distributions while electrons travel randomly through circuits, with photons detected randomly at the end of optical fibres. Many of these events can be described by a Poisson probability distribution.[6]

Consider the specific problem of the random rate of detection of photons from the steady output light of a laser. As each photon is detected, so a single electron is released by the photodetector into a circuit, and in principle the photons can be counted by observing the electron current flow in an interval of time $[t_0, t_0+T]$. With a steady laser output it is found experimentally that, on average, the number of arrivals is independent of the time t_0, but is directly proportional to the interval T.

When the average observations are independent of the starting time t_0, as here, then the process is referred to as a *stationary random process.* If the average rate of detection of the photons is p, then in the interval T we expect pT photons on average, p being known as the *mean count rate.* In particular, for a short enough interval δT, the probability that two photons are detected is proportional to $(\delta T)^2$, and in the limit as $\delta T \to 0$ this is negligible. It is also observed experimentally, that, on average, the photon detection rate is not affected by previous or later intervals of time. So that the fact that one has had less than average numbers of photons arriving in one second does not mean that in the next second more than average numbers of photons will arrive. Events in adjacent intervals are independent.

We now enquire about the probability, $P(n, t, t+T)$, of detecting exactly n photons in the interval t to $t+T$. The stationary property of the process ensures that the time t does not have to be considered, so, dropping t,

$$P(n, t, t+T) = P(n, 0, T) \to P(n, T) \tag{1.5.1}$$

Let us start with $n=0$, so that there are no detections in the interval. The probability $P(0, T+\delta T)$ may be written as the joint probability of no detections in the following interval δT and in the previous interval T. The probability of no detections in δT is $P(0, \delta T)$, so the required joint probability, given no detections in the interval T, is the product $P(0, T)P(0, \delta T)$, using the independence of events in adjacent intervals. Consequently

$$P(0, T+\delta T) = P(0, T)P(0, \delta T) \tag{1.5.2}$$

However, the probability of detecting one photon in the interval δT is $p\delta T$, with a probability of detecting more than one photon being negligible. The probability of no detection of photons in the interval δT is then

$$P(0, \delta T) = 1 - p\delta T \tag{1.5.3}$$

The rate of change of probability can then be discovered from (1.5.2) and (1.5.3) as

$$[P(0, T+\delta T) - P(0, T)]/\delta T = -pP(0, T)$$

In the limit as $\delta T \to 0$

$$d[P(0, T)]/dT = -pP(0, T) \tag{1.5.4}$$

Using the certainty ($P = 1$) of no detections in a zero time interval ($T = 0$), the solution to (1.5.4) gives

$$P(0, T) = \exp(-pT) \tag{1.5.5}$$

The rate of change for $P(n, t, t+T)$, but with $n > 0$, can be similarly found, except that now there are two ways in which $P(n, t, t+T)$ can change in a short additional interval δT. There can be either no collisions in δT and n in the previous T, or alternatively one in δT and $n-1$ in T (more than one collision in δT being neglected). Again ignoring the reference time t, it is possible to link $P(n, T)$ and $P(n-1, T)$ using these alternatives to give

$$\begin{aligned} P(n, T+\delta T) &= P(n, T)P(0, \delta T) + P(n-1, T)P(1, \delta T) \\ &= P(n, T)(1 - p\delta T) + P(n-1, T)p\delta T \end{aligned}$$

In the limit as $\delta T \to 0$

$$dP(n, T)/dT = -pP(n, T) + pP(n-1, T) \tag{1.5.6}$$

The solution where $P(n, 0) = 0$ ($n \geq 1$) is given by

$$P(n, T) = p \exp(-pT)\left[\int_0^T P(n-1, s) \exp ps \, ds\right] \tag{1.5.7}$$

The value of $P(1, T)$ can be found from the known $P(0, T)$ and then $P(2, T)$ can be found from $P(1, T)$, and so on. The general solution is found to be (Fig. 1.5):

$$P(n, T) = [(pT)^n/n!] \exp(-pT) \tag{1.5.8}$$

The Poisson distribution is an important distribution in information processing and analysis, and it is met again in the analysis of a laser and in the next section applied to the arrival of information.

1.6 Information arrival

Telephone calls are paid for by their duration, and although it takes some people a long time to give a one minute message the average rule must be that the amount of information that can be sent is proportional to the transmission time. One difficulty with information is that the information content is subjective. Editors of papers make their reputations by the choice of news that they think is important. Such subjective judgment is unsatisfactory for quantifying information.[7] In trying to find a mathematical definition of 'information' I, a first simple assumption might be that the rate of information arrival is given by

$$dI/dt = \alpha \tag{1.6.1}$$

where α is dependent upon the system used to transmit and receive information, and is determined by the 'information that could be sent' rather than dependent on a subjective view about the value of the information.

To explore this further, suppose that all messages have an equal content of information and are arriving randomly with a Poisson distribution. Suppose that the probability of a message arriving in a small interval $\mathrm{d}T$ or $\beta \mathrm{d}T$, so that the probability of the message not arriving in an interval 0 to T is $P(0, T)$, where P is as defined in the last section with

$$\mathrm{d}P/\mathrm{d}t = -\beta P \tag{1.6.2}$$

or

$$\mathrm{d}[\ln(1/P)]/\mathrm{d}t = \beta \tag{1.6.3}$$

One may think of $\ln(1/P)$ as defining a quantity which may be called 'uncertainty'. The longer the message takes to arrive, then the greater the uncertainty becomes. How well one knows that feeling when waiting for examination results! The message on arriving removes the uncertainty. This suggests that information is related to the uncertainty that may be removed by the message, so it is possible to write

$$I = k \ln(1/P) \tag{1.6.4}$$

where, comparing (1.6.1) and (1.6.4), $k = \alpha/\beta$.

Fig. 1.5. Poisson distribution. $P(n, T) = [(pT)^n \exp(-pT)]/n!$ is the probability of n events in interval T seconds given an average probability of 1 event every $(1/p)$ seconds. $P(n, T)$ has maximum at $pT = n$, and for large enough n the distribution approximates to a gaussian distribution centred on $pT = n$ with standard deviation $n^{1/2}$.

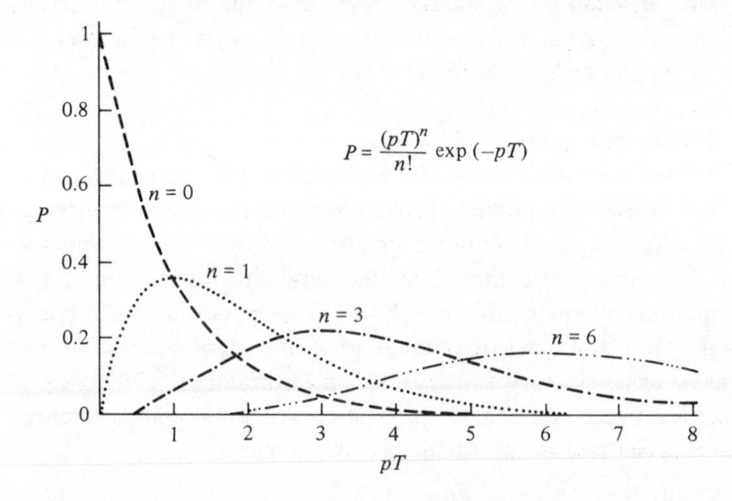

In a physical theory it is not enough to note a mathematical connection which may be quite fortuitous. One should consider the implications and check that they make sense. For example, consider two messages with probabilities of p_1 and p_2, respectively, which form a single message with probability $p_3 = p_2 p_1$. Using the definition of 'uncertainty' as $\ln(1/P)$, the 'uncertainties' removed by these two messages taken together add:

$$\ln(1/p_3) = \ln(1/p_1) + \ln(1/p_2)$$

However, the information in the messages would also be expected to add. Both these results correspond with intuitive notions about information. Information should be additive, and information which may remove a lot of uncertainty tends to be important.

Suppose further that the message was made up from an alphabet of N different symbols which can occur with equal probability, and each message was M symbols long. Then, making use of the concept that information must be a quantity that can be added together to make more information, we expect the possible rate of change of information content with M to be:

$$\partial I/\partial M = \alpha_N \tag{1.6.5}$$

The constant α_N will depend on the 'alphabet' from which the M symbols are drawn. Each letter, having equal probability by the initial assumptions, then contributes $1/N$ to the information. On increasing the alphabet by 1 in an M-lettered message, we expect to find the rate of increase of information with N as given by

$$\partial I/\partial N = gM/N \tag{1.6.6}$$

where g is some constant of proportionality. Taking (1.6.5) and (1.6.6) together leads to $I = gM \ln N = g \ln(N^M)$ with $\alpha_N = g \ln N$. It may also be noted that N^M is the number of ways of arranging M letters from the N-lettered alphabet, so that $1/N^M$ is proportional to the probability of receiving one of the M-lettered messages. The notion that information is given by the 'uncertainty' is again justified.

Finally we observe that the probability for a single member of an N-member alphabet appearing is $1/N$; all members are of equal significance. So $\ln N$ is the information content of the single letter. Introducing the concept of probable information content h, one can see that this is given by

$$h = (1/N) \ln N \tag{1.6.7}$$

The quantity h is sometimes referred to as the 'entropy' of a signal, but if this term is used one must realise that there is no thermodynamic significance to this form of 'entropy'.

The value of h is maximised when $dh/dN = 0$, which occurs when $N = e$, or $\ln N = 1$. There can only be a whole number of symbols so that N has to be integer, and the probable information per symbol is maximised by having $N = 2$ or 3. The value $N = 2$ is preferred because in electronics it is relatively easy to have signals which are either '1' or '0' (on or off) by switching between different electrical voltages. Another important reason for using binary encoded information comes when signals are sent in the form of pulses with known amplitude and duration (e.g. a 1 V pulse, 10 ns duration) as the symbol 1, and no pulse as the symbol 0. Provided that one can detect *any* pulse *above* the background noise then it can be regenerated and reshaped by electronic circuitry into a pulse of the known correct amplitude and known correct duration without amplifying the background noise. Such practical considerations lead to digital systems with binary encoded information. For such systems it is sensible that (1.6.4) should have the arbitrary constant k defined so that

$$I = \log_2 (1/P) \tag{1.6.8}$$

to give a convenient definition of information in terms of removing uncertainty.

PROBLEMS 1

1.1 A two lane motorway has a batch of 50 cars joining the slow lane at one entry point. The statistics show that on average the slow lane cars move to the fast lane at a rate of 20% of cars per mile, while the fast lane cars move back into the slow lane at a rate of 5% of cars per mile.

If the number of cars in the fast/slow lane is F/S, then set up two rate equations for F and S checking that $d(F+S)/dx = 0$ so that cars are conserved. Find the equilibrium numbers for F and S. Estimate after what distance one might expect the numbers in the fast lane to have settled to within 90% of their equilibrium values.

1.2 Given initial concentrations $[H_2] = a$, $[I_2] = b$, with $b > a$ and $[HI] = x = 0$, then show from (1.3.7) that one expects at a later time

$$dx/dt = K_-[R(a - x/2)(b - x/2) - x^2]$$

where $R = K_+/K_-$.

Given that $R > 4$, show that

$$dx/dt = K(A - x)(B - x)$$

where $K = K_{-}(R-4)/4$, $A = g(p-q)$, $B = g(p+q)$, with $g = R/(R-4)$, $p = a+b$, $q = [(b-a)^2 + 16ab/R]^{1/2}$. Hence show that $(A-x)/(B-x) = (A/B)\exp[-K(B-A)t]$. Show that the greatest possible value of x is $2a$.

1.3 A small estate agency finds that it can operate with only two partners buying and selling houses between two very similar areas of property. The partners find that they can each stimulate the same number of sales as spontaneously occurs. Assuming that the model of (1.4.6) held, how many more partners might they be able to take on if they were successful in increasing their sales to 1.25 times the spontaneously occurring sales?

1.4 For a time interval T, the Poisson probability of finding n particles is given by the rate of change $\mathrm{d}P(n, T)/\mathrm{d}T = -pP(n, T) + pP(n-1, T)$. Show directly that $\mathrm{d}\langle n\rangle/\mathrm{d}T = p$ and $\mathrm{d}\langle n(n-1)\rangle/\mathrm{d}T = 2p\langle n\rangle$. Hence, on integration show that $\langle n\rangle = pT$; $\langle n^2\rangle - \langle n\rangle = (pT)^2$. Rearrange to show that $\langle(n-\langle n\rangle)^2\rangle = pT$. ($\langle F_n\rangle$ is the mean value of F_n, i.e. $\langle F_n\rangle = \sum_{n=0}^{\infty} F_n P(n, T)$)

1.5 In a certain production line, the production of a quantity Q of transistors is governed by

$$\mathrm{d}Q/\mathrm{d}t = -bQ + I(t)$$

where the input $I(t)$ fluctuates on a time scale that is long compared to the time $1/b$. Show that one expects approximately

$$Q = I(t)/b$$

This type of argument is frequently used and helps to eliminate additional rate equations (see next problem).

1.6 Two linked processes produce outputs Q_1 and Q_2 such that

$$\mathrm{d}Q_1/\mathrm{d}t = -k_1Q_1 + k_2Q_2 + I(t); \qquad \mathrm{d}Q_2/\mathrm{d}t = k_3Q_1 - k_4Q_2$$

Given that $k_4 \gg k_3$ and that $I(t)$ varies slowly enough, then show that one can neglect the second equation and write

$$\mathrm{d}Q_1/\mathrm{d}t = -k_1'Q_1 + I(t)$$

where $k_1' = k_1 - k_2k_3/k_4$. The second equation is then completely negligible if $k_4k_1 \gg k_2k_3$. If one process happens at an exceedingly fast rate compared to the rates of change of input, one can usually approximate with one fewer rate equation.

Elementary rate equations in semiconductors

2.1 Charge carrier transport

2.1.1 *Introduction*

The motion of charge carriers within a semiconductor[8-10] is governed by a number of useful concepts which can be understood in straightforward ways from rates of change of particles in space, time, momentum and energy. In chapter 1 it was seen that the conservation of particles leads to the continuity equations (1.2.3) and (1.2.5) relating the rate of accumulation of charge density $\partial\rho/\partial t$ and the spatial rate of dispersal div $\boldsymbol{J}$ of current density $\boldsymbol{J}$.

This section shows how rates of change of momentum and energy determine the velocity $v = \boldsymbol{J}/\rho$ of charge carriers as the electric field changes within a semiconductor. In chapter 4 a more detailed approach to carrier transport is discussed using the Boltzmann collision equation, which brings in diffusion and also develops a model relevant to the Gunn effect.

The later sections of this present chapter continue with elementary transport discussing the rates at which semiconductors relax back to equilibrium. Chapter 3 outlines how such rates place limits on the engineering of devices for very high speed switching.

2.1.2 *Rates of change of momentum: mobility*

Quantum theory assures one that electrons behave as waves and that electron waves can travel freely through a *perfectly* periodic array of atoms such as is formed by a perfect crystal. The analogy is often made between electron quantum waves in a crystal and electromagnetic waves travelling through a periodic structure of inductors and capacitors. In such a filter only certain frequencies are permitted to propagate. In the crystal there is equally a limited range of frequencies for the quantum waves, and this means a limited range of energies for the mobile electrons (Fig. 2.1). Ideally there can be no collisions with a host atom in the perfectly periodic array, and so an electric field accelerates any valence electron which is able to move into empty higher energies. However, it is found that on average $v \propto E$ and *not* $\mathrm{d}v/\mathrm{d}t \propto E$ for most practical purposes. This section outlines how rates of momentum change determine this.

Thermal energy vibrates the atoms slightly from their ideal positions in the regularly ordered crystal so that the electric potential of the host atoms in a crystal is no longer perfectly periodic. The electron waves are then scattered slightly by the vibrations and, through quantum and electric forces, exchange energy and momentum with the crystal's vibrational waves.

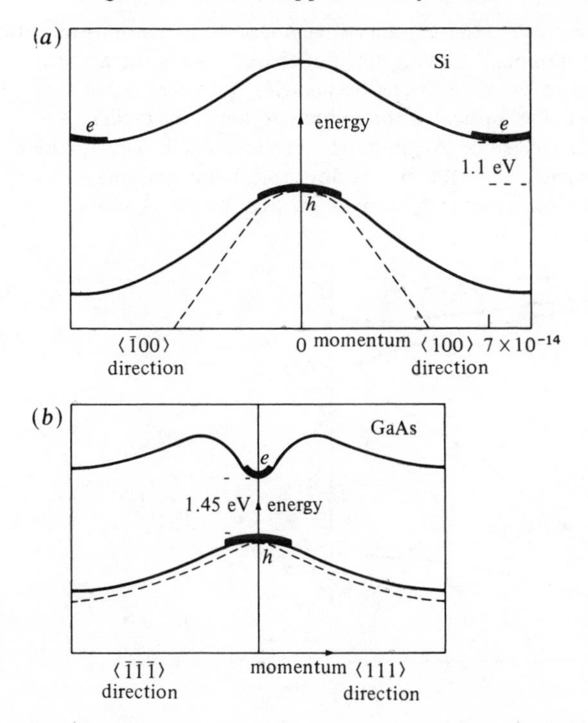

Fig. 2.1. Simplified energy–momentum diagram for the principle permitted states of valence/conduction electrons in a semiconductor. Conduction electrons lie along contour around *e*, valence band holes lie along contour around *h*. (*a*) Indirect gap material (e.g. Si for momentum along 100 axis). (*b*) Indirect gap material (e.g. GaAs for momentum along 111 axis). (Momentum in kg m/s units and both diagrams are drawn approximately to the same scale.)

Quantum theory also assures that any wave has an associated quantum particle, each with energy $\hbar\omega$ and momentum $\hbar k$, where ω is the angular frequency of the vibration and $\lambda = 2\pi/k$ is its wavelength. For vibrational waves in a solid, these quantum particles are known as phonons. Typically the momentum for these particles is $\hbar k \sim \hbar\omega/v_s$, where v_s is the velocity of sound in the material. The electrons then should not be thought of as colliding with the atoms of the crystal but, if a pictorial model is required, colliding with the phonons. Energy and momentum are conserved overall. Thermal vibrations of the lattice take frequencies over a whole spectrum from acoustic frequencies (the same vibrations propagate sound and transmit heat in a crystal) up to a maximum value, typically around 10^{13} Hz (Fig. 2.2). Such high frequency phonons are referred to as *optical* phonons because such frequency values may be associated with infra-red radiation, but of course the optical phonons do not give out light!

The higher the crystal temperature, then, the greater the thermal energy of vibration. In particle terms, the density of phonons is greater, and the rate at which phonons and electrons 'collide' is greater also.

A crystal lattice normally also has defects, where a local change in the perfect periodicity again can scatter the electron wave. Similarly, ionised

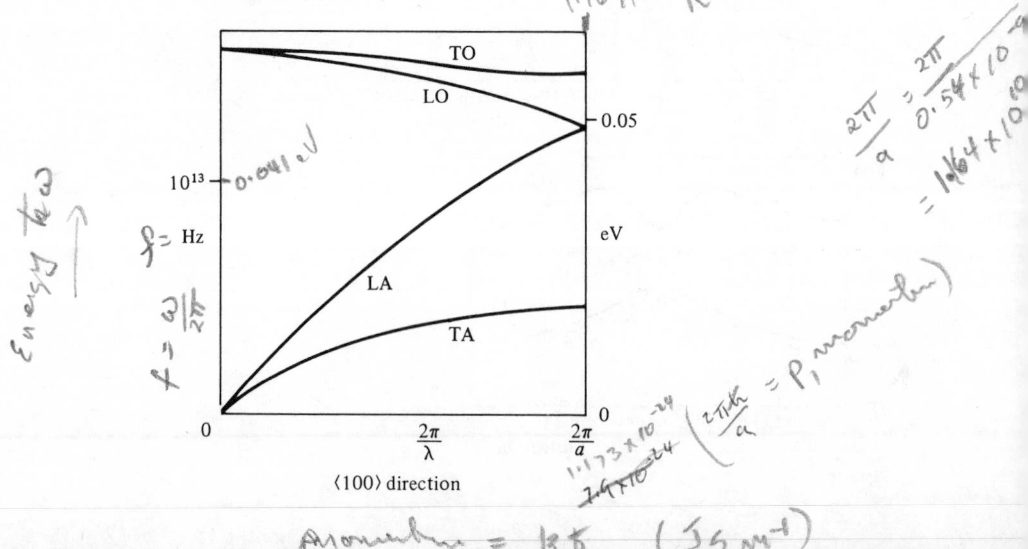

Fig. 2.2. Frequency (energy) and wavenumber (momentum) of lattice vibrations (phonons). Lowest range of frequencies is the acoustic branch - sound waves at lowest frequencies. Upper range of frequencies is the 'optical' phonon branch. Different modes of vibration, transverse or longitudinal vibrations noted with L and T (data represents Si for 100 crystal direction, lattice parameter a = 0.54 nm. Horizontal scale is same as for Fig. 2.1 for phonon momentum).

impurities cause yet further collisions for the electrons. Here the scattering is accomplished directly by electrical fields from the charged ionised impurities acting as deflecting forces on the electrons.

In any one of these 'collision' processes the electron momentum ($\boldsymbol{p}_{\text{electron}}$) changes, losing or gaining momentum to the lattice or impurity (momentum $\boldsymbol{p}_{\text{lattice}}$) at different rates dependent upon the process. One could write formally for one such process (labelled n) that for any one electron

$$\mathrm{d}(\boldsymbol{p}_{\text{electron}})/\mathrm{d}t = \sum_n \{(\boldsymbol{p}_{\text{lattice}}/\tau_{np}) - (\boldsymbol{p}_{\text{electron}}/\tau_{np})\} \quad (2.1.1)$$

where $\boldsymbol{p}_{\text{lattice}}/\tau_{np}$ is the rate at which momentum is given from collisions. The subscript np implies that it is the nth process (i.e. phonon collisions, impurity collisions or defects) and the electron, just prior to the collision, has a momentum $\boldsymbol{p}$. Similarly the electron loses momentum at a rate $-\boldsymbol{p}_{\text{electron}}/\tau_{np}$ because of the same scattering process. The different times τ_n give the effective mean free times between collisions for the different processes. The rate at which the momentum is gained all add together, while the rates at which momentum is lost all subtract.

In equilibrium, with no electric fields, the rate of gain must equal the rate of loss of momentum. Further, with no preferred direction for momentum in the crystal in any collision process, each term averages over all the electrons to zero. With a force present there is a preferred direction and there must be changes.

Consider an applied electric field $\boldsymbol{E}$ giving an average velocity $\boldsymbol{v}$ for each of the conduction electrons; the momentum at any time for a 'typical' electron is

$$\boldsymbol{P}_{\text{electron}} = m^* \boldsymbol{u}_{\text{random}} + m^* \boldsymbol{v} \quad (2.1.2)$$

where m^* is the electron's effective mass in the material. The random velocity $\boldsymbol{u}_{\text{random}}$ has an average magnitude around 10^5 m/s for thermal energies at room temperature and, for values of $\boldsymbol{v}$ much smaller than this thermal velocity, the collision times τ will not change significantly from those with $\boldsymbol{v} = 0$. (2.1.1) then reads for this typical electron

$$\mathrm{d}(\boldsymbol{p}_{\text{electron}})/\mathrm{d}t = \sum_n \{(\boldsymbol{p}_{\text{lattice}}/\tau_{np}) - [m^*(\boldsymbol{v} + \boldsymbol{u}_{\text{random}})/\tau_{np}]\} \quad (2.1.3)$$

The first term and the random component in the second term must still each average to zero as they did in equilibrium, leaving the average rate of loss of momentum for this process:

$$\mathrm{d}(\boldsymbol{p}_{\text{electron}})/\mathrm{d}t = -\sum_n (m^* \boldsymbol{v}/\tau_n) \quad (2.1.4)$$

where τ_n is a characteristic time – the average free time (see problem 2.8) between momentum destroying collisions for the nth type of collision.

The net rate of gain of momentum can only come from the force $e\boldsymbol{E}$, so that averaging over all collision processes yields

$$e\boldsymbol{E} = \mathrm{d}(\boldsymbol{p}_{\mathrm{electron}})/\mathrm{d}t + m^*\boldsymbol{v}/\tau_{\mathrm{m}} \tag{2.1.5}$$

where $1/\tau_{\mathrm{m}} = \sum (1/\tau_n)$ is the 'momentum relaxation time'.

Define a mobility μ_n associated with this nth collision process:

$$\mu_n = e\tau_n/m^*$$

enabling one to define an overall mobility μ from

$$\begin{aligned} \boldsymbol{v} &= \mu \boldsymbol{E} \\ 1/\mu &= \sum_n (1/\mu_n) \end{aligned} \tag{2.1.6}$$

Note that, because collision rates add, so mobilities combine reciprocally. Further, in high quality semiconductors the defects are negligible and the impurity scattering is usually less significant than the phonon scattering.[8]

The 'granularity' of the motion as the electron accelerates, collides, deflects, accelerates again and so on is ignored because the time scales, τ_{m}, of this granularity of motion are found to be so short ($\lesssim 1$ ps) that they are not of current practical interest (i.e. $\mathrm{d}v/\mathrm{d}t \ll v/\tau_{\mathrm{m}}$). New types of semiconductor devices are emerging with submicron dimensions in which the spacings of electrodes are comparable to the mean free paths (~ 10 nm) for electrons between collisions. The 'ballistic' flight between collisions may then be important as indicated in section 4.2.

The 'mobility' concept where $\boldsymbol{v} \propto \boldsymbol{E}$ is valid provided $v \ll u_{\mathrm{random}}$ so that (2.1.3) holds. As v becomes larger, τ_{m} must change from the low field value in order to conserve energy, as considered next.

2.1.3 *Rate of loss of energy*

By balancing the rate of supply of energy with the rate of loss of energy, it is shown here that the electron's velocity cannot continue to increase as the electrical field increases, but must approach a limit – the scattering limited velocity, which we shall use in chapter 3. This limiting velocity arises because energy is exchanged in packets or quanta. There is a maximum frequency, in the 'optical' frequency range of 10^{13} Hz, at which a crystalline lattice can vibrate, leading in turn to a maximum phonon energy $\hbar\omega_{\mathrm{p}}$ per exchange between lattice and electron.

Suppose now that the electric field is sufficiently high so that the electrons always exchange $\hbar\omega_{\mathrm{p}}$ at every electron–lattice 'collision'. Let the characteristic time between such exchanges of energy be defined to be τ_{e} so that the mean rate of loss of energy from an electron is $\hbar\omega/\tau_{\mathrm{e}}$. This loss of energy has to be balanced by the mean rate of supply from the electric field, which is $\boldsymbol{eE}\cdot\boldsymbol{v}$. Hence, for an average electron

$$\boldsymbol{eE}\cdot\boldsymbol{v}\tau_e=\hbar\omega_p/\tau_e \qquad (2.1.7)$$

A very rough guide may take τ_m and τ_e to be similar (typically in the work here the time scales are 1 to $\frac{1}{10}$ ps), but in principle they are different because a collision can easily alter the electron's momentum (a vector quantity) but conserve energy (a scalar), as in an elastic collision. The effects are marked in some semiconductors with $\tau_e>\tau_m$.

Combining the rate of loss of momentum with the rate of loss of energy ((2.1.6) and (2.1.7)) to eliminate the field $\boldsymbol{E}$:

$$\boldsymbol{v}\cdot\boldsymbol{v}=(\hbar\omega_p/m^*)(\tau_m/\tau_e) \qquad (2.1.8)$$

Provided that at high enough electric fields the ratio τ_m/τ_e is independent of the field, the magnitude of the velocity is also independent of the electric field and reaches a scattering limited value:

$$v_s=[(\hbar\omega_p/m^*)(\tau_m/\tau_e)]^{1/2} \qquad (2.1.9)$$

The upper frequency for most lattice vibrations in semiconductors is around 10^{13} Hz, and at high energies the effective masses of electrons within a solid are comparable to the free electron mass. The square root also reduces variations so that it appears difficult to find materials which have scattering limited velocities greatly in excess of 10^5 m/s. A typical velocity/field characteristic for electrons (or holes) within a semiconductor is expected to look as in Fig. 2.3(*b*), though it will be found that

Fig. 2.3. Velocity–electric field characteristics. (*a*) GaAs (electrons), (*b*) Si (electrons) and (*c*) holes (similar for both GaAs and Si). (Note all velocities end close to 10^5 m/s at high fields.) (See chapter 4 for discussion on negative slope at velocity/field.)

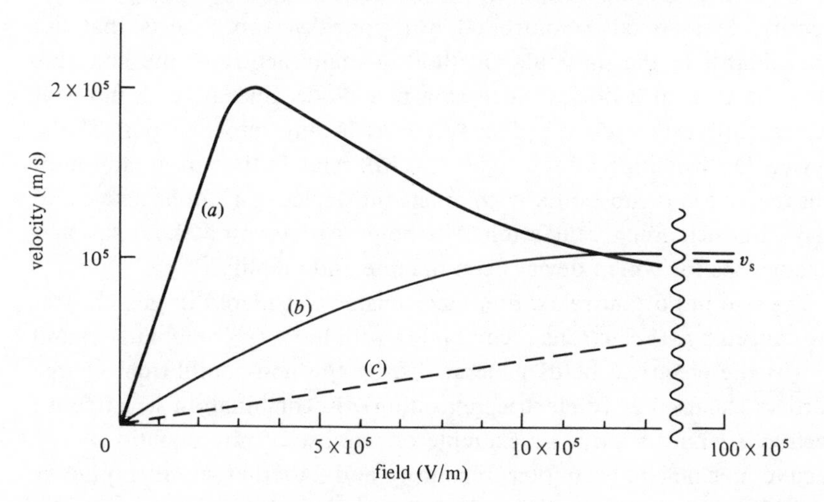

there are significant deviations from this profile with materials such as GaAs (Fig. 2.3(*a*)).

The argument above does not imply that the characteristic time scales of collisions do not change with the electric field. Indeed it is implicit that as the field rises, increasing the rate of delivery of energy, so the rate at which the phonons remove energy must also rise. Once this assumption breaks down, the energy of the electrons will increase until new limits are reached for energy transfer. This leads to impact ionisation of valence electrons which can also be considered by the energy balance.[11] However, in the interests of a compact discussion only a more elementary account of impact ionisation is given (section 3.2).

In conclusion, rates of transfer of momentum and energy between electrons, lattice and field show why $v \propto E$ for low fields and $v \rightarrow v_s$ at high fields.

2.2 Relaxation rates

2.2.1 *Introduction*

In the engineering of semiconductor devices to switch and amplify voltages and currents at high rates, consideration has to be given not just to the velocity of the electrons or holes but also to the rate at which a device (which is in some active state) can relax into a quiescent or equilibrium state. In this chapter we consider two such rates as examples which will be used later in the book.

In any semiconductor material there is some equilibrium charge density n_0 for the mobile conduction electrons and p_0 for the mobile holes in the valence band. The values of n_0 and p_0 are determined by the density of impurities that are deliberately inserted to control the charge carrier density, and are also controlled by impurities and defects that are unavoidable in the particular method of manufacture of the material. Suppose then in a device, such as a p–n diode, one drives a pulse of current injecting a charge density $n > n_0$ locally into one part of the device. On switching off the current, n will relax in time to n_0, and until this relaxation is substantially complete the device will still be active and may conduct. Such relaxation will help to determine the rates and frequencies for which devices can operate and amplify.

The two important relaxation mechanisms considered in this chapter are caused by (i) electrons recombining with holes (recombination) and by (ii) the electrical fields generated from the non-equilibrium charge carriers themselves (dielectric relaxation). Recombination is left until section 2.3 and is a classic example of one use of rate equations. The second mechanism is simpler in concept and underlies so much physics of devices that it is essential to include it in any discussion.

In equilibrium with no current flowing, the positive charge in a device balances the negative charge. For example, in a metal the magnitude of charge of the mobile electrons equals the positive charge on the ionised host atoms which each contribute to the conduction electrons. Similarly, in a semiconductor, the number of conduction holes p (positive charge) minus the number of conduction electrons n (negative charge) together with the charge of the ionised donors or acceptors must, on average, vanish. Suppose that this were not true and there was, for example, a net positive charge. This charge would create an electric field which would attract electrons into the region of positive charge, and the field would persist until the charge was neutralised giving no average field. Certain local variations of this are permitted, especially with the process of diffusion considered in chapter 3, but this principle of electrical neutrality simplifies the physics of many practical devices taken as a whole.

2.2.2 *Dielectric relaxation time*

From electromagnetic theory, the electrical field of any local excess of charge density ρ_e is controlled by Gauss's law:[12]

$$\iint_S \boldsymbol{D} \cdot \mathrm{d}\boldsymbol{S} = \text{net charge } Q \text{ enclosed within surface } S \text{ (volume } V)$$

$$= \iiint_V \rho_e \, \mathrm{d}V \tag{2.2.1}$$

In its differential form, Gauss's law is

$$\operatorname{div} \boldsymbol{D} = \rho_e \tag{2.2.2}$$

where the electric flux density is $\boldsymbol{D} = \varepsilon_r \varepsilon_0 \boldsymbol{E}$, with ε_r as the relative permittivity of the medium. The excess charge density $\rho_e = \rho - \rho_0$, where ρ_0 is the equilibrium charge density.

For materials obeying Ohm's law, the current density $\boldsymbol{J}$ and electrical field $\boldsymbol{E}$ are related by the conductivity σ:

$$\boldsymbol{J} = \sigma \boldsymbol{E} \tag{2.2.3}$$

where the conductivity is proportional to the mobile charge density, $\sigma = \mu \rho_0$. The continuity relationship (1.2.5), with (2.2.2) and approximately uniform σ and ε_r, can be rewritten as

$$(\sigma / \varepsilon_r \varepsilon_0) \operatorname{div} \boldsymbol{D} + \partial \rho / \partial t = 0 \tag{2.2.4}$$

or

$$\rho_e + (\varepsilon_r \varepsilon_0 / \sigma) \, \partial \rho_e / \partial t = 0 \tag{2.2.5}$$

with the solution $\rho_e = \rho_{e0} \exp(-t/\tau_d)$, where $\tau_d = (\varepsilon_r \varepsilon_0 / \sigma)$. Hence, any charge density ρ that is injected suddenly into a device in excess of the equilibrium charge density ρ_0 is neutralised in a time scale of τ_d, the *dielectric relaxation time*, by the motion of mobile charge carriers which contribute to conduction. Practical orders of magnitude for this time scale can be estimated. The permittivity $\varepsilon_r \varepsilon_0$ of many semiconductors is of the order of 10^{-10} F/m. Conductivities range from less than 10^{-5} S/m for high resistivity (semi-insulating) material to more than 10^5 S/m for high conductivity material, giving respective dielectric relaxation times of several microseconds down to femtoseconds. Provided that one is concerned with devices where the changes of current through a device take much longer than a picosecond, the normal practical approximation with medium to high conductivity material is to assert that the material is electrically neutral.

2.2.3 *Motion of charge carriers*

In many practical devices the charge carriers are constrained to move roughly in one direction permitting an approximate one-dimensional analysis. However, as can be seen from Fig. 2.4, even if the charge is constrained to move in one dimension, the electrical field may

Fig. 2.4. Space charge reduction. (*a*) Ideal one-dimensional fields $E = Q/2\varepsilon$ for symmetrical fields from excess charge Q per unit area. (*b*) Even if the motion of charges is confined to one dimension, the electric fields can diverge out in three dimensions reducing the force on neighbouring charges; $F_1 \sim 0.1$. (*c*) The presence of a metal can greatly lower the longitudinal fields through the effect of the image charge causing most of the flux to go transversely; $F_2 \sim 0.01$. (Values are merely rough orders dependent on geometry.)

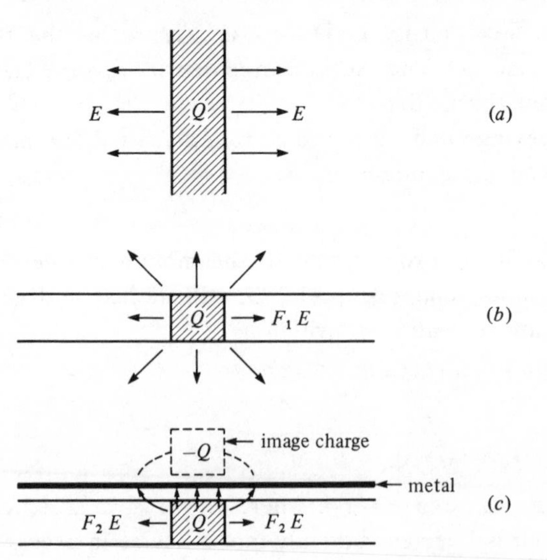

be able to escape all around any excess charge so that the electric flux in the $0x$-direction is greatly reduced from the true one-dimensional value. A useful practical one-dimensional form of Gauss's law then is to write

$$\partial(\varepsilon_r \varepsilon_0 E)/\partial x = F\rho_e \qquad (2.2.6)$$

where F is known as the space charge reduction factor[11,13] and depends upon the geometry. For example, if a metal electrode is brought right up against a thin layer of excess charge (as in an FET; see chapter 3) then the factor F can be easily as low as $1/100$ because the electric flux from the excess charge is forced to go transversely to the closely adjacent image charge in the metal rather than along the $0x$-direction of motion of charge in the layer.

Suppose that an excess carrier density ρ_e is moving in the $0x$-direction with a velocity $v(E)$, which is a function of the electric field E with the current density J given by ρv. The general non-linear analysis is not easy, but if it is supposed that only small changes v_1, ρ_1, E_1 and so on are made on uniform steady state values v_0, ρ_0, E_0 etc., then some results are readily obtained. Neglecting the squares of the small 'signal' values with subscripts 1

$$v_0(\partial\rho_1/\partial x) + \rho_0(\partial v_1/\partial x) + (\partial\rho_1/\partial t) = 0 \qquad (2.2.7)$$

However,

$$v_1 = (\mathrm{d}v/\mathrm{d}E)E_1$$

(2.2.6) yields

$$\partial(\varepsilon_r \varepsilon_0 E_1)/\partial x = F\rho_1 \qquad (2.2.8)$$

Taking $\mathrm{d}v/\mathrm{d}E$ and ε_r as uniform, then (2.2.8) combined with (2.2.7) yields

$$(\partial\rho_1/\partial t) + v_0(\partial\rho_1/\partial x) + [(\mathrm{d}v/\mathrm{d}E)F\rho_0/\varepsilon_r\varepsilon_0]\rho_1 = 0 \qquad (2.2.9)$$

The equation is then interpreted as stating that the charge decays as it travels along with a velocity v_0, and the decay time is the relaxation time constant

$$\tau = \varepsilon_r\varepsilon_0/(\mathrm{d}v/\mathrm{d}E)F\rho_0 \qquad (2.2.10)$$

This is a generalisation of dielectric relaxation and leads to the important result that if $\mathrm{d}v/\mathrm{d}E$ is negative, as it can be in certain semiconductor materials when they are biased by some appropriate electric field, then charge will accumulate at least initially until the increasing field leads to $\mathrm{d}v/\mathrm{d}E = 0$ and no further growth. The result of (2.2.10) will be referred to later in discussions about devices made from GaAs and InP. The inclusion of charge carrier diffusion has further consequences and dampens any growth or decay (see problem 4.6).

2.3 Recombination

2.3.1 *Introduction*

A second major mechanism for restoring charge densities to equilibrium is recombination.[14] There are frequent occasions in semiconductor devices where the overall charge is electrically neutral because of the short time scale of dielectric relaxation. However, one may still have an excess or deficit of an equal number of electrons and holes which neutralise each other. The electrical neutrality means there is no significant electric field associated with these excess charges so that equilibrium has to be reached by another route – 'recombination'.

Recombination is a process whereby electrons at higher energies spontaneously fall to any available vacant energy states at some lower energy. In particular, electrons in the conduction band of a semiconductor fall down to fill vacant sites formed by holes in the valence band. Rate equations help considerably in understanding this process and discovering the role of impurities in the middle of the band gap which can help to encourage recombination.

2.3.2 *Direct recombination*

To start the discussion, consider the simplest process whereby a conduction electron loses energy directly and falls into a vacant hole site in the valence band (Fig. 2.5). Such a direct process occurs only in certain materials where the loss of energy $\mathscr{E}$ ($\sim$the band gap energy) generates a quantum of electromagnetic radiation, a photon, with energy $h\nu = \mathscr{E}$. A more detailed consideration will be given to this direct process in section 6.2.2 on photon–electron interactions. Interactions between electrons and other particles require that the total energy and momentum

Fig. 2.5. Direct recombination. An electron falls from the conduction band (CB) into the valence band (VB) with emission of photon. The momentum of the photon is small, on this scale, compared to that of the electron, and the transition conserves momentum and energy overall.

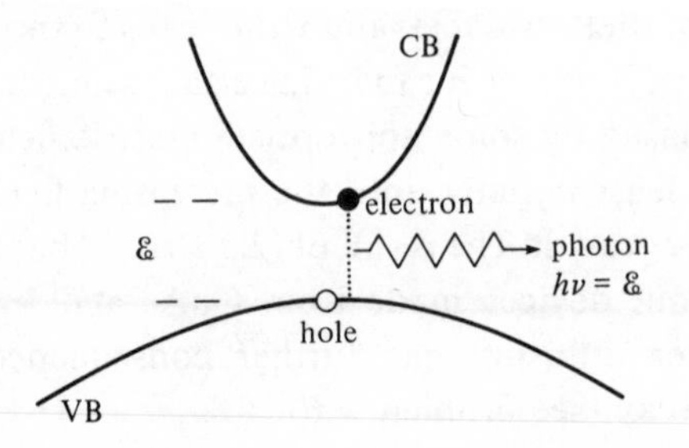

are all conserved. So, in silicon, such direct recombination is not possible because in changing from the lowest part of the conduction band to the highest part of the valence band an electron has to change momentum by a large amount in comparison with the momentum of a photon. An additional contribution to the momentum is required (Fig. 2.6) which can be supplied by the lattice vibrations (phonons) or by having localised impurities which effectively can interact with a wide range of momentum and assist in recombination.†

Returning to direct recombination, consider a unit volume of an appropriate material in which direct recombination can occur. Let there be p holes and n electrons in this unit volume. Each electron has a probability of recombining with any one hole, and this probability in unit time is defined to be R. However, with p holes there is a probability of Rp recombinations for each electron, giving the net rate of loss of electrons as

$$\partial n/\partial t)_{\text{recombination}} = -Rpn \tag{2.3.1}$$

Now it has to be understood that there is occasionally, at non-zero temperatures, a photon of adequate energy to ionise a valence electron and raise it into the conduction band. This thermal generation in the

† A local point disturbance in space can be Fourier analysed into a sum of waves $\iiint \exp(j\boldsymbol{k}\cdot\boldsymbol{r})\,\mathrm{d}k_x\,\mathrm{d}k_y\,\mathrm{d}k_z$ showing that it can in principle interact with waves of any wavelength $(2\pi/k)$ and hence with wave particles of any momentum $\hbar k$.

Fig. 2.6. Indirect recombination. (*a*) With an indirect gap material, the photon with the normal band gap energy cannot supply enough momentum, and electron states at F are normally full so that a direct transition cannot occur. A phonon with a small amount of energy and requisite momentum will help the transition. (*b*) Introduction of impurity levels leads to states with a wide range of momentum, and transitions can more easily occur (see section 2.3.3).

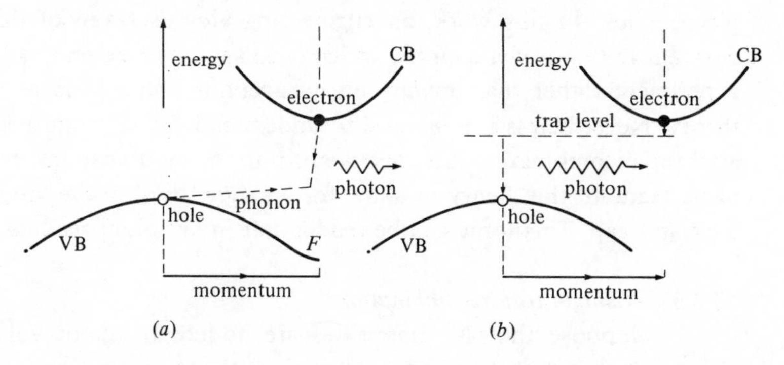

steady state has to balance the recombination so that the equilibrium number of holes and electrons can be maintained. Allowing for this generation G, assuming that the background thermal energy remains constant,

$$\partial n/\partial t)_{\text{net recombination}} = -Rpn + G \tag{2.3.2}$$

Given that n_e and p_e are the equilibrium values of p and n when $\partial n/\partial t = 0$, then $G = Rn_e p_e$ and

$$\partial n/\partial t)_{\text{net recombination}} = -R(pn - p_e n_e) \tag{2.3.3}$$

In chapters 5 and 6, taking into account interactions required to conserve energy, it will be seen that rate equations can reveal the statistical distribution of particles with energy and show that (2.3.3) is consistent with classical statistical distributions, where

$$n = N_c \exp[-(\mathscr{E}_c - \mathscr{E}_f)/kT] \tag{2.3.4}$$

$$p = N_v \exp[(\mathscr{E}_v - \mathscr{E}_f)/kT] \tag{2.3.5}$$

$$np = n_i^2 = N_c N_v \exp[-(\mathscr{E}_c - \mathscr{E}_v)/kT] \tag{2.3.6}$$

for electrons in a conduction band at energy $\mathscr{E}_c$ and valence band at energy $\mathscr{E}_v$ with a reference energy, or 'Fermi level', $\mathscr{E}_f$. The fixed thermal generation requires $p_e n_e = n_i^2$, independent of $\mathscr{E}_f$, in the material.

Consider, for example, a p-type material with an impurity density N_A, where the acceptors are nearly all ionised so that all the impurities contribute a hole, $p \approx p_e \approx N_A$ and $p \gg n$. One then has approximately

$$\partial n/\partial t)_{\text{net recombination}} = (n - n_e)/\tau_{rn} \tag{2.3.7}$$

where $\tau_{rn} = 1/RN_A$. The recombination is proportional to the excess carrier density $(n - n_e)$ with a recombination rate that increases as N_A increases.

This linearised model for recombination is used time and again because of its simplicity and help in indicating the correct approximate physical behaviour. Values of recombination time vary from seconds down to picoseconds. In this work, an engineering view is taken of this time constant in that it is a quantity which is measured or determined by the experiment rather than relying on an accurate value from a detailed theory. Nevertheless it is helpful to understand the mechanisms which alter the recombination rate, and we turn to the modifications that have to be made to the theory to allow for a single 'trap' in the 'middle' of the band gap. This topic can be readily left in an initial reading.

2.3.3 *A single trap recombination*

Suppose that N_{tr} impurities are added to a unit volume of semiconductor such that these isolated impurities trap electrons from the conduction band by inserting energy levels in the 'middle' of the band gap (Fig. 2.6(*b*), Fig. 2.7). As indicated earlier, it is relatively easy to transfer momentum from and to isolated traps so that immediately one suspects that recombination should be assisted. Let the trap levels be at an energy $\mathscr{E}_{tr}$, with $\mathscr{E}_c > \mathscr{E}_{tr} > \mathscr{E}_v$. Further, it is supposed that in this example these traps are neutral in charge unless an electron is present. The traps may be referred to as 'deep' level acceptors because they behave like the normal 'shallow' acceptors, but the words deep and shallow refer to the positions in the energy band gap in comparison with the energy band edges.

There will then be n_{tr} electrons trapped on these impurity sites with $(N_{tr} - n_{tr})$ vacant traps available to accept further electrons. Because the Fermi energy $\mathscr{E}_f$ may be close to the energy $\mathscr{E}_{tr}$, the full Fermi–Dirac probability function (see section 5.2) must be used without the usual simplifying Boltzmann approximation. The expected occupation of the traps will be:

$$n_{tr}/N_{tr} = 1/\{1 + \exp[(\mathscr{E}_{tr} - \mathscr{E}_f)/kT]\} \tag{2.3.8}$$

Consider then n electrons in the conduction band and assume that the only significant way that these electrons can be removed is by falling in energy into one of the $(N_{tr} - n_{tr})$ vacant sites. The argument is just like the initial argument, so that

$$\partial n/\partial t)_{\text{recombination}} = -R_n[n(N_{tr} - n_{tr})] \tag{2.3.9}$$

where R_n is the rate constant to be determined from experiment or from more detailed theory. As previously, there has to be a rate of thermal generation of the n_{tr} trapped electrons in proportion to their numbers, driving them into the conduction band. This generation rate is defined then as $R_n n_{e1} n_{tr}$ with n_{e1} to be found. Using the same arguments as in

Fig. 2.7. Energy level diagram for recombination with traps. Traps (deep level acceptors) trap electrons from the conduction band. They can enhance recombination and lead to compensated semi-insulating material.

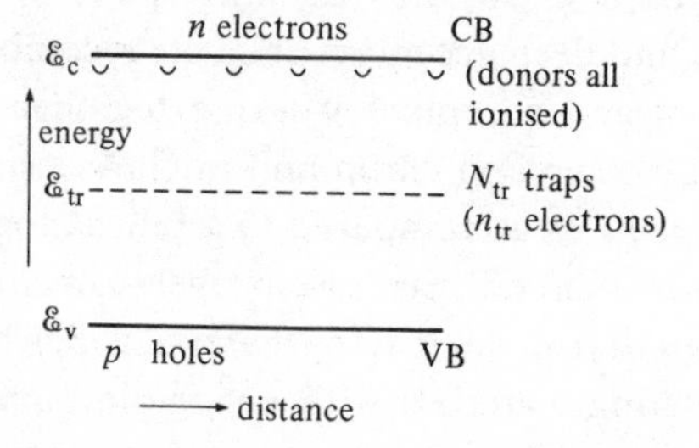

section 2.3.2,

$$\partial n/\partial t)_{\text{net}} = -R_n\{[n(N_{\text{tr}} - n_{\text{tr}})] - [n_{\text{e1}} n_{\text{tr}}]\} \tag{2.3.10}$$

The unknown density n_{e1} is determined from the equilibrium conditions.

Suppose, for example, that the Fermi energy $\mathscr{E}_{\text{f}}$ came exactly at $\mathscr{E}_{\text{tr}}$, then $N_{\text{tr}} - n_{\text{tr}} = n_{\text{tr}} = \frac{1}{2}N_{\text{tr}}$. At this level it is expected that as many electrons recombine as are thermally generated and (2.3.10) must give zero net recombination. It follows that n_{e1} is equal to the density of electrons in the conduction band if the Fermi energy is at the trap energy level.

The arguments now all follow through, making the necessary changes, for holes p in the valence band, so that

$$\partial p/\partial t)_{\text{net}} = -R_{\text{p}}[pn_{\text{tr}} - p_{\text{e1}}(N_{\text{tr}} - n_{\text{tr}})] \tag{2.3.11}$$

with p_{e1} equal to the density of holes in the valence band if the Fermi energy was at the trap energy level.

Now combine the results of (2.3.10) and (2.3.11) to discover the change of trap occupation:

$$\partial n_{\text{tr}}/\partial t)_{\text{recombination}} = R_{\text{n}}\{[n(N_{\text{tr}} - n_{\text{tr}})] - [n_{\text{e1}} n_{\text{tr}}]\} - R_{\text{p}}\{pn_{\text{tr}} - [p_{\text{e1}}(N_{\text{tr}} - n_{\text{tr}})]\} \tag{2.3.12}$$

When the traps have stabilised, $\partial n_{\text{tr}}/\partial t)_{\text{recombination}} = 0$, then

$$n_{\text{tr}} = N_{\text{tr}}(R_{\text{n}} n + R_{\text{p}} p_{\text{e1}})/[R_{\text{n}}(n + n_{\text{e1}}) + R_{\text{p}}(p + p_{\text{e1}})] \tag{2.3.13}$$

Inserting this value of n_{tr} into (2.3.10),

$$\partial n/\partial t)_{\text{net recombination}} = (np - n_{\text{e1}} p_{\text{e1}})/(\tau_{\text{tr}} N_{\text{tr}}) \tag{2.3.14}$$

with

$$\tau_{\text{tr}} N_{\text{tr}} = [R_{\text{n}}(n + n_{\text{e1}}) + R_{\text{p}}(p + p_{\text{e1}})]/(R_{\text{n}} R_{\text{p}} N_{\text{tr}}) \tag{2.3.15}$$

It follows immediately that the higher the trap density, the shorter the recombination time. For a p-type material with $p \approx N_{\text{A}}$, then p_{e1} would be negligible in comparison to N_{A} as would n and n_{e}, so that similar to (2.3.7)

$$\partial n/\partial t)_{\text{net recombination}} = (n - n_{\text{e}})/\tau_{\text{rn}} \tag{2.3.16}$$

with $\tau_{\text{rn}} \approx 1/R_{\text{n}} N_{\text{tr}}$.

It is important to realise that it is not only deep *acceptors* that assist recombination. Deep donors and deep acceptors enhance recombination in similar ways. Both donors and acceptors working together can be particularly effective as with oxygen and chromium in GaAs, where the recombination times can be ~0.1 ns as compared to a few nanoseconds for direct recombination in pure material. Such deep levels can effectively remove mobile carriers in short time scales leaving material which behaves like an insulator (semi-insulating material) with negligible numbers of charge carriers. Problem 2.7 examines a normally n-type material with large numbers of acceptor traps and shows that the electrons must be removed by the traps. The Fermi level at equilibrium would be just below the trap level, which can be similar to that level for ideally pure material.

Gold in silicon can similarly enhance the rate of recombination (τ_r changing from microseconds to nanoseconds) through introducing 'deep' energy states. Deep states can be produced in otherwise pure material through bombarding the material with high energy (~100 keV) protons, which create dislocations of the lattice. These dislocations insert energy states in the middle of the band gap because the periodicity is no longer perfect in the crystal. Surfaces also provide 'dangling valence electron bonds' which again introduce energy levels in the middle of the band gap, so enhancing recombination.

There are devices, such as charge coupled devices,[15] where recombination and generation must be minimised to ensure that a packet of excess charge remains constant as it moves through a semiconductor. It follows that pure indirect band gap materials are likely to succeed best in this application. Silicon is the obvious choice when long recombination times (hundreds of microseconds to milliseconds) are required.

PROBLEMS 2

2.1 If the mean free time τ_m is independent of the energy of the conduction electrons, explain why collisions amongst electrons themselves should not change the mobility. Will this conclusion still be necessarily true if τ_m is dependent upon energy?

2.2 Estimate the energy required by an electron to ionise a valence electron in a semiconductor and so produce an electron–hole pair. Assume a band gap energy $\mathscr{E}_g$ and equal electron and hole effective masses. Remember that momentum and energy are conserved.[11]

2.3 A set of impurity levels in a semiconductor traps electrons of density n at a rate n/τ. The concept of a capture cross-section σ is often used. If n electrons travel with an average (thermal) velocity v ($\frac{1}{2}m^* v^2 = 3kT/2$), then they each sweep out a capture volume $v\sigma$ per unit time. If there are N traps/unit volume explain why $1/\tau = Nv\sigma$. Estimate σ for a trap rate of 1 ns given 10^{23} traps/m^3. Take $kT = 25$ meV and $m^* = 10^{-30}$ kg.

2.4 It is observed for some very pure materials that the mobility varies with temperature T as $\mu \propto T^{-3/2}$. Extend the concept of capture cross-section to a collision cross-section σ_{pe} for collisions between phonons and electrons and suggest the implications of this result for the density of phonons.

2.5 Two terminals of arbitrary geometry have a uniform material between them of conductivity σ and permittivity ε. The resistance between the two electrodes is R and the capacitance is C. Show that $RC = \tau_d$, the dielectric relaxation time given by ε/σ. It may be assumed that the electric field is confined wholly within the material.

(Hint: divide geometry into 'cubes' with walls parallel to field lines and equipotentials noting that field lines are normal to equipotential surfaces. For small enough elements cube distortion may be neglected – use curvilinear orthogonal coordinates.)

2.6 Using the arguments of section 1.2, show that with recombination the one-dimensional continuity equation must read $(\partial\rho/\partial t)+(\partial J/\partial z) = -(\rho - \rho_e)/\tau_r$, where $1/\tau_r$ gives the recombination rate. Integrate over distance to show that with a single current input $J_{in} = (Q/\tau_r)+(\partial Q/\partial t)$, regardless of the spatial distribution provided τ_r is independent of position $(\partial\rho_e/\partial t = 0)$.

2.7 In a particular material with acceptor traps (as for (2.3.12)) it is known that N_{tr} is much larger than the number of donors. There are negligible numbers of holes because without the traps the material is n-type. Explain why there can be a negligible density n of conduction electrons in equilibrium.

2.8 Suppose that an electron has its collisions distributed with a Poisson distribution so that in any small time interval δT the probability of a collision is $\delta T/\tau$. Show, by considering the probability of no collisions in time T, that the probability of a collision in δT at T after the last collision is $[\exp(-T/\tau)](\delta T/\tau)$. Hence show that, in a field E, the average value of increase of momentum between collisions, is

$$\langle m^*(v_{final} - v_{initial})\rangle = eE\tau$$

Given that $v_{initial} = 0$ always, after any collision, show that the average distance travelled between collisions is $eE\tau^2/m^*$. Hence show that the average velocity is $\langle v\rangle = eE\tau/m^*$; and observe that $\tau = \tau_m$, the mean time between momentum destroying collisions.

3

Rates of switching

3.1 The simple switch: an introduction

In the hope of avoiding analysis for its own sake, some of the ideas of the preceding chapter are applied to the dynamics of semiconductor switching devices which are important in computer and communication systems. Most texts on semiconductor devices concentrate on impurities, Fermi levels, diffusion equations and equilibrium starting conditions. In the approach here, the chief concerns are the rates at which a device can transport charge. So RC time constants, the dielectric relaxation rate, transit times and rates of recombination are the quantities that appear in approximate dynamic analyses of the selected switching devices. There is an inevitable tendency to digress from one or two themes of rate equations into standard semiconductor physics, and the forbearance of the reader is requested when the digressions into physics are too lengthy, inadequate or both.

Digital communications and computations rely on transmitting information by electromagnetic pulses (electrical, microwave or optical) which are either 'on' or 'off'. In section 1.6 it was seen that such binary signals helped to maximise the probable information of a single symbol. Moreover, encoding signals into pulses leads to more accurate detection, regeneration and transmission of data through a variety of techniques such as error correcting codes, which can combat interference or noise in a transmission path. Faster switching rates lead to shorter pulses and so to higher rates of data and information processing.

An ideal switch between a load and a source would transfer power instantaneously to the load, but we remind the reader here how switches and loads have, of necessity, lead and contact resistances with stray capacitances which limit the rate of transfer of charge to the load.

Figure 3.1 shows some first-order practical effects, where the switch has an 'on' resistance R_s (leads and contacts) between a source (voltage V and internal resistance R_i) and load resistance R_L. Practical loads always have some shunt capacitance C_L. If the switch is suddenly turned on, the simplest calculation considers the rate of increase of the charge Q to the load from the current flow:[16]

$$Q = C_L V_L \tag{3.1.1}$$

$$\mathrm{d}Q/\mathrm{d}t = [(V - V_L)/(R_i + R_s)] - (V_L/R_L) \tag{3.1.2}$$

Writing $\tau = (R_s + R_i)C_L$; $a = (R_L + R_i + R_s)/R_L$ then

$$\tau\, \mathrm{d}Q/\mathrm{d}t + aQ = C_L V \tag{3.1.3}$$

With the initial condition, $Q_{\text{initial}} = 0$

$$Q = (VC_L/a)[(1 - \exp(-at/\tau)] \tag{3.1.4}$$

In computers with a capacitive metal-oxide semiconductor device for a memory which stores charge, R_L will be large so that $a \to 1$. The 'on' resistance (including any line loss from the switch) together with the device capacitance, not allowing for the actual switching time, gives a rise time (defined as time taken for the voltage to go from 10% to 90% of its total change in value) of $\tau_{rs} = 2.2(R_s C_L)$. Further delays increase the rise time (problem 3.8).

It is often argued that the microelectronic technique of reducing the size of everything reduces these parasitic RC time constants. So taking

Fig. 3.1. The switch. Source of voltage V and internal resistance R_i, with ideal switch to load resistance R_L. Contact/lead resistor R_s of switch. Load has capacitance C_L, charge $Q = Q_0[1 - \exp(-t/\tau_{rs})]$. Rise time for charge or voltage is $2.2\tau_{rs}$.

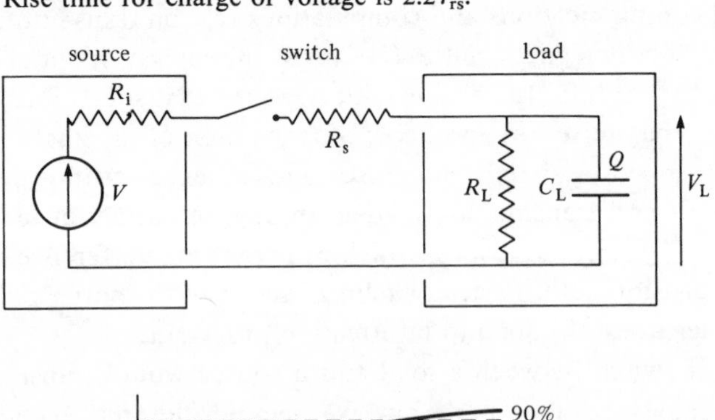

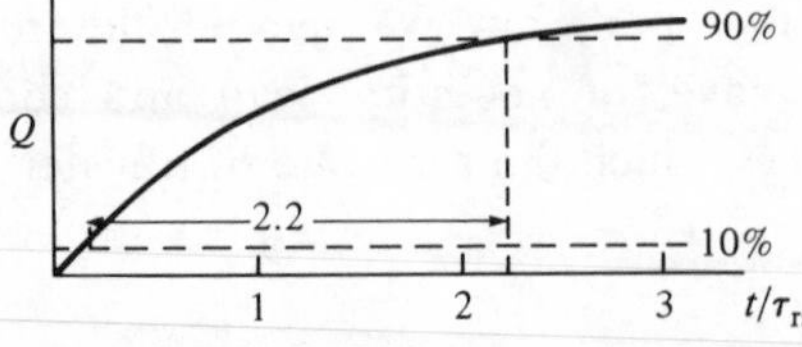

an ideal geometry as in Fig. 3.2

$$R_s C_L = (\varepsilon/\sigma)(AL/Wt\delta) \tag{3.1.5}$$

where σ is the conductivity of the metal line and ε is the permittivity for the material of the capacitor. If the metallisation remains the same thickness δ then indeed, if all other dimensions are reduced by a factor k, $R_s C_L$ reduces by k. However, the metallisation of the leads cannot be kept a constant thickness once the dimensions of the lines reach micron dimensions. The thickness δ has to be scaled with the line width W ($\delta \ll W$) and, worse, the resistivity of thin layers of metal rises away from the ideal bulk value. Scaling to too small dimensions limits the lead/load rise time (problem 3.2) and even can increase the RC time constants!

Designers for subnanosecond switching have to appreciate the importance of such RC limitations even though these seem trivial at first sight. Inductances in the leads in general add further delays. Although there are ways of reducing the effect of stray capacitances and inductances by 'embedding' the high speed switching device into a transmission line or electrical filter circuit (see problem 3.3) these give small improvements.

The main point from this discussion is that the rate of transfer of charge is often a key limitation to the speed of switching. Interconnection RC effects will be one limit on the speeds at which large computers can operate. Contact resistances, lead inductances and additional stray capacitances can all further reduce efficiency and speed.[17]

3.2 Types of switch

3.2.1 *The plasma switch*

It is useful to categorise switches into two broad classes: 'plasma switches' and 'charge transport switches'.[18] Although some devices mix the effects together, usually one or the other principle is predominant. The essential differences may be understood by contrasting a 'simple'

Fig. 3.2. Scaling for microcircuits. Metalisation thickness δ cannot remain constant but must eventually also scale as dimensions are reduced. Limits on $R_s C_L$ time constant – see problem 3.2.

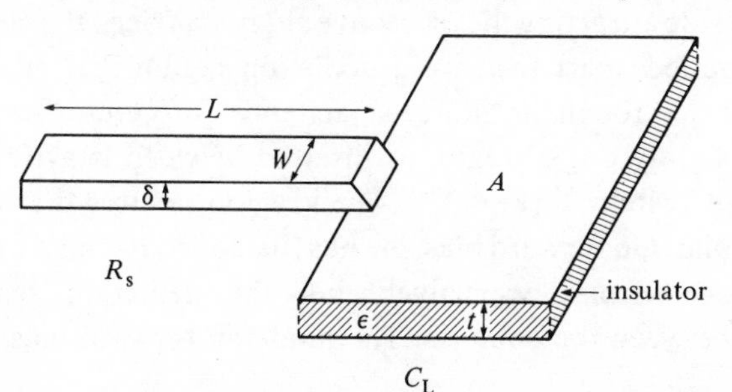

resistance R_s and a conducting diode which takes a current I and voltage V such that $V/I = R_s$ as for the resistor. The following discussion develops this.

A 'plasma' switch is defined as any switching device where the conducting path, when the switch is 'on', is formed from an electrically neutral plasma of positive and negative charges which conduct the current. Thus, in a simple resistance R_s, conduction electrons are, on average, neutralised by a background positive charge of ionised host atoms donating the conduction electrons. Alternatively a semiconductor device might be filled with equal numbers of holes and electrons. Then any excess electrons, which are injected at one contact, create a local excess electrical field which propagates through the conductor with the speed of electromagnetic waves, forcing electrons at the other end to leave so as to preserve internal charge neutrality, as in the discussion on dielectric relaxation. With microcircuit engineering reducing dimensions so that electromagnetic propagation delays are negligible, it is the dielectric relaxation rate which limits the frequencies that can be transmitted through the switch.

The plasma switch has two different limitations: first the rate at which the plasma can be established; secondly, the frequencies for the electromagnetic signals which can be handled by the switch. The latter depends to a first order on the capacitance and resistance of the plasma switch. The former depends on how the plasma is created. Effects of inductances are ignored in this first-order discussion, though with large devices they may be important.

3.2.2 *Methods of plasma formation*

Three methods of plasma formation are discussed briefly here: (i) plasma injection from suitable contacts, (ii) photoionisation, and (iii) impact ionisation of valence electrons.

(i) For a p–n junction diode, the conventional switching action occurs as the diode switches from normal reverse bias (off) into forward bias (on). In reverse bias, with a positive voltage on the n-side and a negative voltage on the p-side attracting the respective charge carriers, the electrons and holes are pulled apart to leave a depletion region that effectively conducts only r.f. through its finite capacitance. In certain switching diodes, a short lightly doped region is inserted between heavily doped contacts: a p–i–n switch[1,2] (Fig. 3.3). The i-region reduces the capacitance on reverse bias. On forward bias, the heavily doped p- and n-contacts inject holes and electrons, respectively, into the 'depletion' region of semiconductor between the contacts. The diode on forward bias can be regarded as a plasma switch capable of conducting even microwave signals. However, (section 3.4) the plasma is established and removed at relatively slow rates determined by recombination.

Thyristors[1,2,20] are other p–n semiconductor devices where a gate triggers the injection of a plasma of holes and electrons to give a high conductance switching path. These devices are useful for electrical power control, for lighting and machines, and can switch kilovolts and kiloamps in submicrosecond time scales.

(ii) A hole–electron plasma can also be created by photons with sufficient energy (greater than the band gap or binding energy for valence electrons) to photo-ionise the valence electrons into the conduction band and leave equal numbers of holes in the valence band, automatically generating an electrically neutral plasma (Fig. 3.4).[1,2,12,18,23]

The rate of production of the photogenerated plasma is limited by the rate at which the light may be absorbed within the semiconductor material (discussed in chapter 6). It is not difficult to find combinations of photon energy and semiconductor material such that the absorption is accomplished within a micron from the surface, equivalent to a femtosecond or so for absorption time. Problem 3.6 shows that if a picosecond pulse of light with photon energy $\mathscr{E}_p$ and total energy $\mathscr{E}$ is absorbed within a gap length L between two contacts to semiconductor with mobilities $\mu_{n/p}$ for the electrons/holes, then the order of the switch resistance between two contacts is given by

$$R_s = L^2\mathscr{E}_p/[q\mathscr{E}(\mu_n + \mu_p)] \tag{3.2.1}$$

Putting in numbers indicates that resistance of a few ohms can be obtained with picojoules of light given micron gap dimensions.

Fig. 3.3. p–i–n diode. (*a*) Reverse bias, holes and electrons pulled apart – device capacitance C. (*b*) Forward bias with i-region filled with conducting plasma of holes and electrons – resistance R_s.

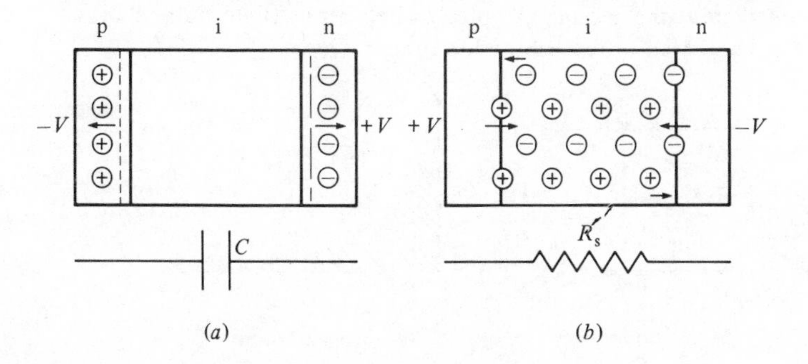

The conductivity of the material decays after the light pulse has ceased because of recombination. Slow enough rates of recombination can maintain the switch in the 'on' state, while faster recombination rates are required to turn the switch off rapidly. Practical optoelectronic switches using these principles can switch a hundred or so volts in picosecond time scales.[18,21]

(iii) Any particle with enough energy can in principle ionise an atom through a collision and so, in a crystal, produce a new electron in the conduction band along with a hole in the valence band.[1,2,22,23] A significant proportion of the mobile electrons or holes in a semiconductor can gain sufficient energy from a strong enough electric field to ionise, on the next 'impact', any atom in the crystal, releasing more mobile electrons and holes to produce an avalanche (Fig. 3.5) of charge carriers. The material can be transformed from high resistivity with only a few carriers to a dense highly conducting plasma.

Typically this change is performed with a semiconductor p–n junction where, with large enough reverse bias voltages, the electrical field in the depletion region becomes sufficiently high to initiate strong impact ionisation. The charge carriers have mean free path $L \sim 10$ nm. Now to reach an ionisation energy $\mathscr{E}_i > \mathscr{E}_{\text{band gap}}$ the electrical field E must do work on the electron over a distance L_i, where $L_i = \mathscr{E}_i / qE$, where motion only in the direction of the field is considered. From discussions about Poisson distributions (section 1.5) the proportion of carriers travelling the distance L_i is $\exp(-\mathscr{E}_i/qEL)$. So the probability of an ionisation is $A(E) \approx p(E) \exp(-\mathscr{E}_i/qEL)$, where $p(E)$ is the probability of an ionisation once the electrons have reached the required energy.

The rate at which impact ionisation generates new electrons, from the present number n, will be strongly dependent upon the field E and is

Fig. 3.4. Photo-ionisation. (*a*) Principles – electron–hole pairs created if $h\nu > \mathscr{E}_c - \mathscr{E}_v$. (*b*) Conductivity δ with depth. Light shining on surface is absorbed exponentially with depth ($\alpha \sim 10^6$ m) creating a conducting plasma of holes and electrons (light pulse short compared to recombination time).

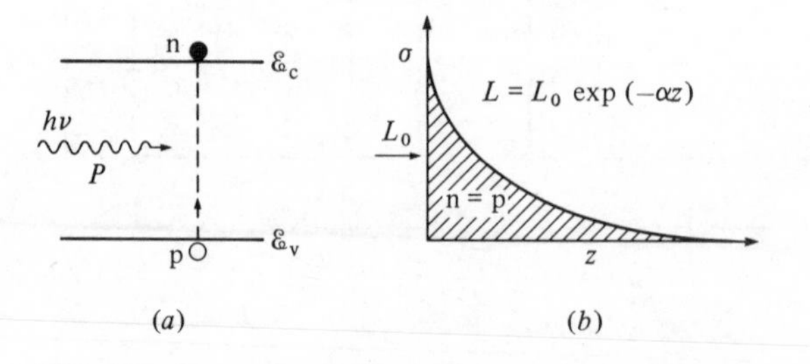

given from

$$dn/dt = A(E)n/\tau \tag{3.2.2}$$

where τ is the mean free time ~1/10 ps.† With strong enough fields the depletion region can be filled with a plasma in picosecond time scales.

This method of switching is used in the avalanche mode for switching transistors and in avalanche microwave oscillator diodes.[1,22]

In areas outside semiconductors, the mercury wetted reed relay with a suitable circuit is a surprising mechanical switch which can operate in picosecond time scales.[18] This speed cannot be created by the velocity at which the reeds close, but must occur through rapid ionisation of the mercury vapour forming a plasma around the contacts. The hydrogen thyratron is a gas filled triode where the switching is initiated by ionising a gas to form a highly conducting plasma of electrons and positive ions. This is a robust device which can switch many kiloamps and kilovolts in submicrosecond time scales.

† It is more usual to write $dJ_{n/p}/dz = \alpha_{n/p}J_{n/p}$ and to distinguish between the ionisation rates $\alpha_{n/p}$ of holes and electrons, but this lies outside the scope of this discussion.[1,22,23]

Fig. 3.5. Schematic diagram of avalanche production of plasma by impact ionisation. Single electron gains energy from strong field E to ionise atom and produce electron and hole which, in turn, produce more electrons and holes. The figure shows the example of four collisions with four more holes and four more electrons produced. Note that the current at the left equals the current at the right.)

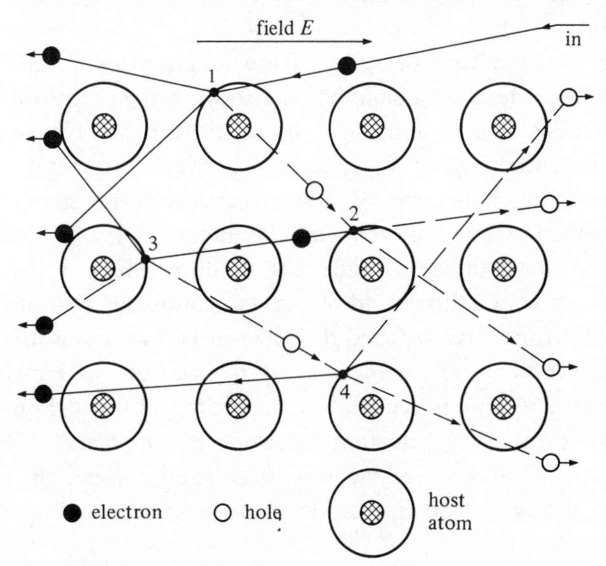

3.2.3 *Switching rates*

Impact ionisation and photo-ionisation can produce a plasma so rapidly that one of the main limitations on switching speed is the finite plasma resistance R_s and capacitance C_d between the switch's contacts. Changes i in the switch's current I and changes v in the voltage V across the switch are controlled approximately by the equation

$$i = (v/R_s) + C_d(\partial v/\partial t) \tag{3.2.3}$$

The time constant R_sC_d is at best just the dielectric relaxation time, and may be so short that the switch is limited by R_sC_L as in section 3.1. With large devices propagation times and inductances may also need to be considered.

Plasma switches using p–n junction diodes may be formed by simply switching the diode from forward to reverse bias and vice-versa (see section 3.4). The time to change the state of the switch is determined by the recombination time, but once the switch is 'on' the diode may pass high frequencies (even microwaves) as indicated in (3.2.3) with suitable R_s, C_d values.

3.2.4 *The charge transport switch*

In the plasma switch an electron entering at the cathode creates a field forcing some other electron to leave at the anode. In contrast, a charge transport switch is one where the charge carriers have to move right across the device from one contact to the other for conduction to occur. One of the simplest examples of such a switch is a Schottky barrier diode (SBD).[1,2,24,25]

Figure 3.6 gives an electron energy band diagram for the SBD. The cathode of the diode is an ohmic contact to an n-type semiconductor, while the anode is a suitable metal evaporated over the layer of n-type material. A potential barrier Φ_b is formed between the metal and the semiconductor. This barrier prevents the electrons in the metal moving into the conduction band of the semiconductor. When the anode is positive with respect to the cathode, the diode conducts by electrons flowing from the conduction band of the semiconductor into the metal. Once these electrons have reached the metal it is found that it requires only subpicosecond times for them to be neutralised (in a metal, the dielectric relaxation time is negligibly short for most practical purposes) and travel as ordinary conduction electrons in the metal. (The fact that the electrons arrive in the metal with an energy above the normal thermal energy has caused this diode to be known as a 'hot carrier' diode.)

On reverse bias, with the metal negative with respect to the n-semiconductor, the potential barrier Φ_b prevents significant electrons flowing from the metal. The voltage also attracts electrons in the semiconductor away from the metal, so widening a region which is depleted of conduction electrons. The depletion region acts as a 'parallel plate' capacitance. Any direct current on reverse bias is limited by leakage and the thermal generation rate. With good technology the leakage is negligible, and so is the thermal generation for many applications.

The diode then can act as a simple switch - conducting on forward bias, but with only a capacitance on reverse bias. However, even if the diode has the same 'on' resistance ($V/I = R_s$) as a plasma switch there is still a vitally important difference. The electrical fields in the diode cannot force electrons into the 'anode' unless they have first been emitted at the 'cathode'. Conduction in the diode then requires the electrons to pass right across the depletion region at the interface between the cathode (n-type material) and the anode (the metal). The transit time, τ, taken for the carriers to cross this region is given approximately by 10 ps per micron of material. This time is about three orders of magnitude longer than the propagation of electromagnetic waves through a plasma in a semiconductor and so can provide a major additional limitation to the speed of switching, even with the micron dimensions made possible by modern technology.

Fig. 3.6. Energy band diagram for Schottky barrier diode. (a) Equilibrium - potential barrier between conduction electrons in metal and conduction electrons in semiconductor. (b) Forward bias - electrons move from semiconductor into metal. Transit time $\tau \sim$ 10 ps/μm distance. (c) Reverse bias voltage taken up by wider depletion region as electrons are attached by bias away from metal. Barrier Φ_b prevents carriers moving from metal into the semiconductor.

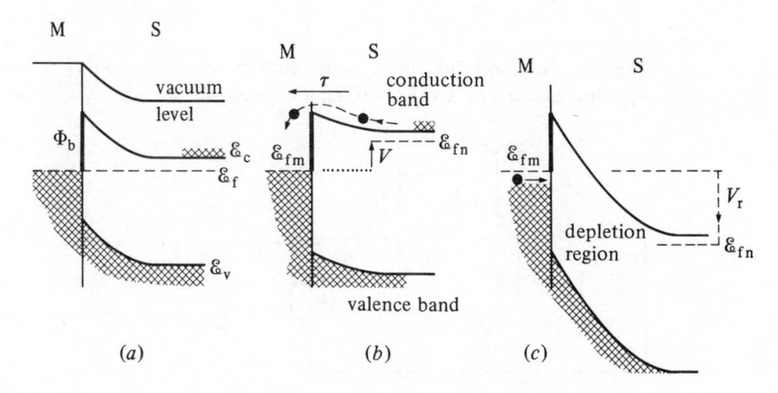

3.2.5 *First-order modelling of charge transport switch*

Consider then a current I flowing through the diode, with the charge carriers taking a time τ to flow from anode to cathode. The magnitude of electronic charge Q stored between the cathode and anode must be

$$Q = I\tau \tag{3.2.4}$$

It must now be recognised that there is an average delay of approximately $\frac{1}{2}\tau$ in the build up of this travelling stored charge. The rate of supply of charge at the cathode, where the field is zero, is I. The rate of loss of charge at the anode is approximately $Q(t-\frac{1}{2}\tau)/\tau$. The difference between these two currents represents the rate of increase of charge in between the contacts:

$$\mathrm{d}Q/\mathrm{d}t = I - [Q(t-\tfrac{1}{2}\tau)/\tau] \tag{3.2.5}$$

Expanding to a first order in the transit time τ gives the approximate dynamic equation

$$I \sim \tfrac{1}{2}(\mathrm{d}Q/\mathrm{d}t) + (Q/\tau) \tag{3.2.6}$$

Further insight into this result can be obtained through another argument. The total current flowing into any device is composed of convection current (given by the motion of charge carriers including any diffusion of these carriers) and Maxwell's displacement current (density $\partial D/\partial t$ with $D = \varepsilon E$). Consider a diode of area A and length L from cathode to anode, containing a charge density $\rho = -qn$ moving with a velocity v, with the total current I_t found by assuming a one-dimensional flow (Fig. 3.7). Taking positive current flow from left to right:

$$-I_t = -Aqnv + A\varepsilon\, \partial E/\partial t \tag{3.2.7}$$

Integrating over the length from anode to cathode gives the voltage

$$V = \int (-E)\,\mathrm{d}x \tag{3.2.8}$$

Fig. 3.7. Induced current. Charge element $-qn\delta x$, moving at velocity v, induces current $I_t = \int qn(v/L)\,\mathrm{d}x$ in terminals with V constant.

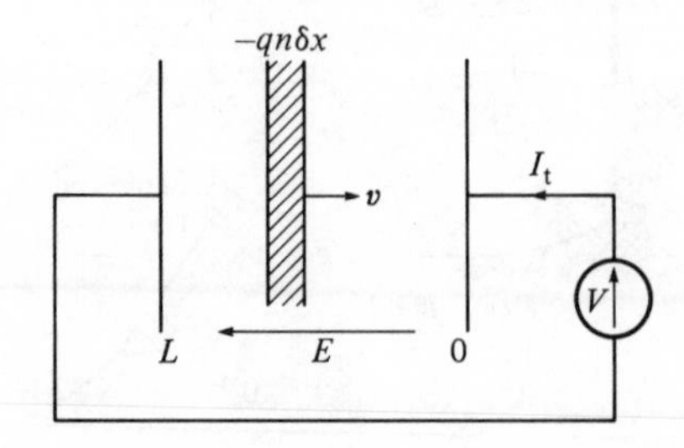

The total current I_t has to be the same throughout the length of the device (from current conservation or Kirchhoff's law), so integrating (3.2.6) over the whole length yields

$$I_t = \int (dQ/\tau) + C_d(\partial V/\partial t) \tag{3.2.9}$$

where $dQ = Aqn\,dx$; $\tau = L/v$ is the local transit time across the diode; and $C_d = A\varepsilon/L$ is the diode's capacitance in the absence of charge (Fig. 3.7). The term dQ/τ is called the *induced* current for the charge element dQ; the moving charge can be in the middle or anywhere in the device and still induce current flow.

If the charge was travelling with a uniform velocity v then the density n would vary in space-time as a function $f(x - vt)$, so that near the anode $n = f(L - vt - y)$, with the distance y measured back from the anode. For a steady state condition, the charge flowing into the anode would be $Q/\tau = \int qf(L)v\,dx/L$. With time varying conditions, $n \sim f - y\,\partial f/\partial y = f + (y/v)\,\partial f/\partial t$ with $f = n$ at the anode still. On integrating in (3.2.9), taking the average delay again as $\frac{1}{2}\tau$,

$$I_t \sim (Q/\tau) + \tfrac{1}{2}(\partial Q/\partial t) + C_d(\partial V/\partial t) \tag{3.2.10}$$

(3.2.10) then gives the same result as (3.2.6) except that we have now included the capacitive current $C_d\,\partial V/\partial t$ that has to be present even if no convective currents flow.

For the convective currents, it is known for semiconductor diodes that in the steady state ($\partial/\partial t = 0$):

$$I = I_0[\exp(qV_0/\eta kT) - 1] \tag{3.2.11}$$

η is 1 for the ideal device.

To demonstrate the implications, assume that the diode capacitance and carrier transit time τ, on forward bias, are approximately constant. (3.2.6) may then be considered with small changes of current i, charge q and voltage v. Define a charge storage capacitance C_s from

$$q = i\tau/2 = vC_s \tag{3.2.12}$$

From the steady state values for small changes $v/i = R_s = kT/\eta qI_0$ so that $C_s = \tau/2R_s$. Considering the dynamic addition of dq/dt to the steady current, one has the first-order switching equation

$$iR_s = v + \tau_t(\partial v/\partial t) \tag{3.2.13}$$

where

$$\tau_t = R_s(C_d + C_s) = R_sC_d + \tau/2$$

The result of (3.2.13) for a charge transport switch looks formally similar to that for the plasma switch, except that the time constant is increased by the transit time.

Figure 3.8 indicates features for a fast Schottky barrier diode. The active diode depletion region is kept thin, made perhaps by a 0.1 μm layer of epitaxial material grown over a highly doped material as a contact, to keep any series resistance low. The transit time through the depletion layer can then be around 1 ps and the depletion capacitance with very small areas ($<50\ (\mu m)^2$) can be kept to below 0.1 pF. Such diodes have been made to operate at frequencies up to and above 100 GHz.[25]

3.3 The field effect transistor

3.3.1 *Introduction*

Field effect transistors are outstandingly good examples[26,34] of fast charge transport switches which rely on charge being transported right across the device from source to drain. Unlike the Schottky barrier diode in the preceding section these devices have a gate to control the switching.

There are two basic types of FET: devices which are normally off and devices which are normally on when the control gate voltage $V_g = 0$. The former are often preferred in logic or computing systems because they consume less power, generate less heat and so can have a higher packing

Fig. 3.8. Section through fast Schotty barrier diode. Thin n-type layer to keep transit time short. Highly doped rear contact. Whole device mounted on insulator to keep stray capacitances to ground low (insulator can be semi-insulating GaAs for example). Top contact can be 'transparent' to permit light to enter n-region for fast photodetection.

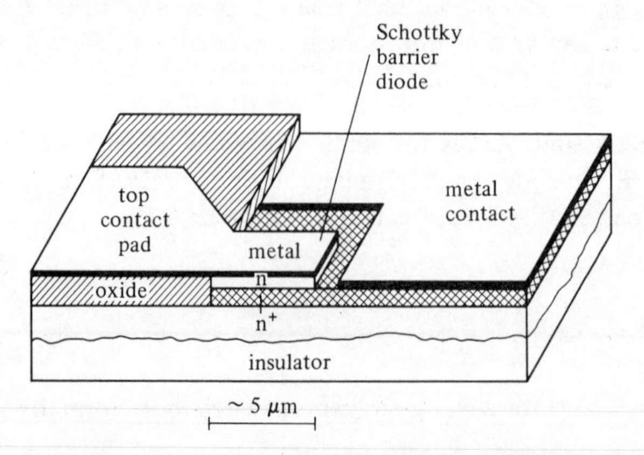

density. However, it would stray too far to go into the device physics of all the different types so here we look at one type of field effect transistor using a general dynamical model which considers rate of change of charge; a technique that is applicable to a variety of devices.

3.3.2 *Operation and construction*

Figure 3.9 shows an electron energy diagram illustrating one particular principle for the operation of practical switching and amplifying devices. There is a source, or an emitter, of conduction electrons with a drain, or collector, at a higher voltage (lower potential energy for electrons) than the source. The flow of electrons from the source to the drain is controlled by a gate which controls a 'hump' in the potential energy. The electrons must surmount this barrier before they can reach the drain (or collector). Once the electrons have enough energy to overcome the hump they must fall down the potential slope to the positively biased collector. This model illustrates in a qualitative manner the principles for field effect transistors, newer 'camel' transistors using p–n junctions,[22] the permeable base transistors[34] and also the high electron mobility transistors.[25] However, the need to be explicit will limit the discussion to a Schottky barrier field effect transistor - SBFET - (Fig. 3.10).

The SBFET is also referred to as the MESFET (MEtal–Semiconductor FET) and is currently made using the GaAs rather than Si because conduction electrons in GaAs have a higher mobility than conduction electrons in Si. As will be seen later, the faster the electrons move for a given electrical field then the higher the switching speed. Moreover,

Fig. 3.9. Generalised band diagram for switching device. Gate controls potential hump that in turn controls numbers of charge carriers (electrons) able to surmount hump and reach collector (drain). Once electrons have surmounted the hump the electrical fields force the electrons to the collector (drain).

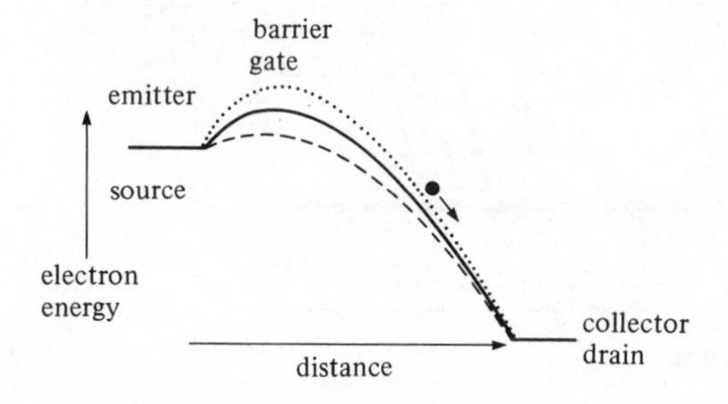

GaAs/metal junctions form Schottky barriers, with appropriate characteristics, more easily than Si/metal junctions. The semiconductor InP turns out to be another good material to use for MESFETs.

Referring to Fig. 3.10 a conducting channel of n-type material is grown on a buffer layer of pure GaAs, which acts like an insulator (around 10 MΩ cm resistivity). Ohmic contacts to this channel form the source and the drain. The gate is a Schottky barrier which, on reverse bias with the metal negative with respect to the source, repels the electrons beneath the gate, depleting the channel of charge carriers. When this conducting channel is fully depleted, the electrons are forced to move in a narrow layer close to the interface of the undoped pure high resistivity GaAs substrate. The gate then provides a potential hump for the electrons flowing from the source to the drain. The more negative the gate then the fewer electrons can reach the drain.

A theory due to Shockley[28] can be found in many texts[1,2,26] where steady state I/V characteristics are determined by the change of conductance in the channel between source and drain as the electrons are repelled from the channel beneath the negatively biased gate. The dynamic model given here complements this static model.

3.3.3 *Dynamics*

Here we are concerned with the dynamics of the device, and to this end the problem is simplified to contain only its basic elements (Fig.

Fig. 3.10. A schematic diagram of a Schottky barrier field effect transistor. Two devices in parallel – sources are connected to same point externally. Source – drain dimensions L of order 1 μm, gate width $W \sim 100$ μm, channel thickness $\sim\frac{1}{4}$ μm.

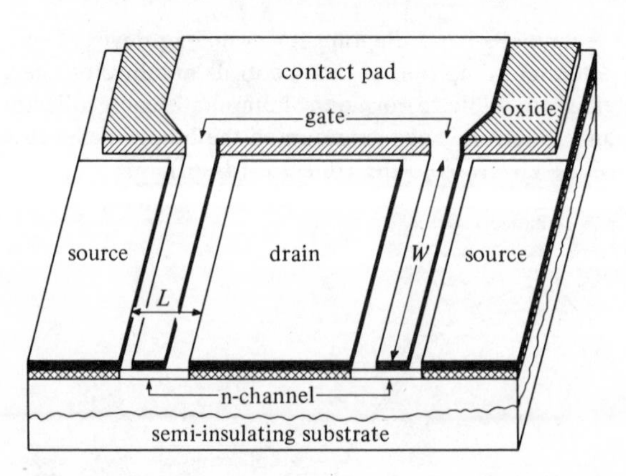

3.11). The gate is biased negatively repelling the electrons beneath the gate and so forming a depletion region which creates a capacitance C_{gc}, the gate-channel capacitance. We shall assume that this capacitance is roughly constant, which is consistent with the conducting channel being fully depleted, so that the charge carriers flow close to the interface between the channel and insulating substrate. The electrons, moving at close to scattering limited velocities, take approximately a constant time τ_g to cross this region as they travel from source to drain. Once the electrons have crossed the potential barrier of the gate they are forced by the electrical fields to the positively biased drain regardless of the exact potential of the drain. This corresponds to a current generator I_{ds} for current flowing away from drain towards the source (electrons flowing towards the drain). In the steady state we have to have the charge storage relation (as discussed for the diode):

$$I_{ds} = Q/\tau_g \qquad (3.3.1)$$

This travelling charge was stored before the current to the drain could be fully established, leading to a delay between Q and I_{ds} of approximately $\frac{1}{2}\tau_g$. Dynamically, a change i_{ds}, for a change q in the charge, is expected to be:

$$i_{ds}(t) \equiv q(t - \tfrac{1}{2}\tau_g)/\tau_g$$

or

$$i_{ds}(t + \tfrac{1}{2}\tau_g) \equiv q(t)/\tau_g \qquad (3.3.2)$$

For changes varying as $\exp j\omega t$, retaining first-order terms:

$$i_{ds}(1 + \tfrac{1}{2}j\omega\tau_g) = q/\tau_g \qquad (3.3.3)$$

The current through the device is established by the charge under the gate creating fields driving the electrons to the collector. The time delay is seen to be in the correct sense. Now if the change in voltage across

Fig. 3.11. Circuit elements for dynamic model of FET. r_c = source-gate channel resistance; C_{gc} = gate depletion capacitance; r_g = gate resistance.

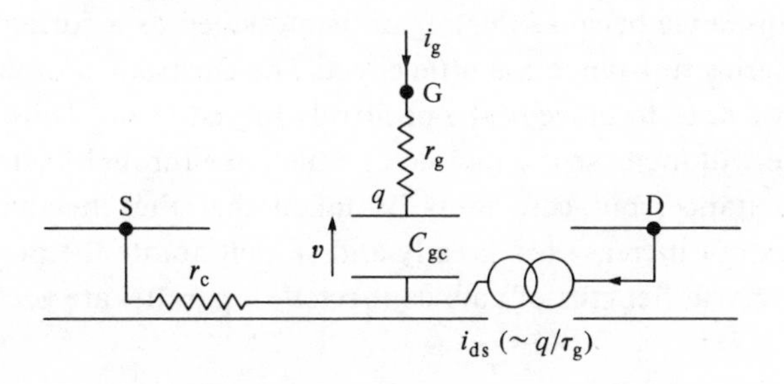

the gate-channel capacitance is v then

$$q = vC_{gc} \tag{3.3.4}$$

However, the rate at which this charge changes is determined by the gate current

$$i_g = dq/dt = v(j\omega C_{gc})$$

The current through r_c is $i_g + i_{ds}$, while the current through r_g is i_g, so that

$$v_{gs} = v[1 + j\omega C_{gc}(r_c + r_g)] + i_{ds}r_c \tag{3.3.5}$$

Retaining again the first-order terms in ω, one arrives at the result that

$$i_{ds} = [g_m/(1 + j\omega\tau)]v_{gs} \tag{3.3.6}$$

where the mutual conductance is

$$g_m = (C_{gc}/\tau_g)/[1 + (r_c C_{gc}/\tau_g)]$$

and

$$\tau = \tfrac{1}{2}\tau_g[1 + g_m(r_c + 2r_g)]$$

3.3.4 *Consequences of the dynamical model*

The reader is assumed to have used the concept of mutual conductance [$i_{ds} = g_m(\omega)v_{gs}$] and to understand that power gain and switching speed depend upon controlling large drain currents with small gate-source voltages.[32] The results of (3.3.6) will help point the way to understanding the design and performance of present and future devices. First, $g_m(\omega)$ decreases with frequency at a rate which depends on the effective transit time through the device. This effective time is increased by the time taken to charge the gate capacitance through the gate resistance r_g and the source-gate channel resistance r_c. Second, the magnitude of the mutual conductance also depends on the transit time; the shorter the transit time, then the larger the value of g_m. The value of g_m is reduced by any channel reistance. Practical FETs may have the gate offset towards the source (Fig. 3.10) so as to reduce the source-gate resistance and also reduce the strong fields which can lead to breakdown between gate and drain. Extra resistance in the gate-drain channel has less effect on the dynamic performance because this region is modelled as a current generator where series resistance has little effect. The electrons once having crossed the gate *have* to move to the positively biased drain. Third, long lengths of gate will increase the mutual conductance through increasing the gate capacitance, but care must be taken that the effective gate resistance does not increase too greatly and so deteriorate the performance. Several shorter lengths of gate connected in parallel are preferred (Fig. 3.10).

This model, using rates of change of the stored charge, does not rely on too much of the detailed physics and so applies to other devices. For example, it can apply to the family of insulated gate field effect transistors. The parameters such as the transit time and the additional stray capacitances will be different. Insulated gate FETs rely, for example, on attracting electrons to the surface of a p-type material (inversion) and so opening a conducting channel between n-type source and drain contacts. Now the gate has to extend right across the region between source and drain to ensure a complete conducting channel. The gate has then significant capacitances between the source and drain which have not been included in the present simplified model. Having to charge and discharge additional capacitances invariably reduces the speed of switching. Insulated gate FETs are not as fast as SBFETs because of this factor and because of material problems and surface states which permit insulated gate FETs to be made well only in Si which has a lower mobility than GaAs.

The model used here also explains the rationale of some future developments. Material with higher mobilities have a higher channel conductance reducing the parasitic channel resistance r_c. New structures are being engineered by advanced methods such as epitaxial growth using molecular beams,[29,34] forming materials which have a higher limiting velocity for the conduction electrons and so reduce the transit times. Smaller dimensions for both gates and source–drain spacings can be defined using finely focused electron beams rather than normal photolithography. Submicron dimensions are being used again to reduce the transit time and capacitances but in favourable proportions. Indeed, with small enough dimensions between the contacts, it is believed that the electrons will have negligible numbers of collisions which remove energy from the electrons so that they will move almost as in a vacuum between the contacts. The generality of this type of charge storage model though should hold for even these smaller devices and show how rate of change of charge can help understand the way forward, at least until one reaches regions where quantum theory and the statistical behaviour of few electrons starts to affect the answers. However, technology and material science will play major roles in determining such limits.

3.3.5 *Circuit effects on switching speeds*

In practical high speed logic circuits one typically needs to have the output from one device fanning out to drive several other devices. A fan out of three or more is considered essential. However, if the input impedance of the device is too low or the output impedance too high then insufficient power can be transferred to effect the switching of future devices. Even if enough power can be transferred, the speed is invariably reduced by the load. It is essential to consider the additional circuit loading to determine the overall performance. No particular new principles about rates of switching are involved, but the technology of making high power FETs consistent with these principles is a taxing problem.

3.4 Notes on charge storage models

3.4.1 *Introduction*

To illustrate further the importance of charge transfer in switching devices, this section gives notes on first-order dynamic models of p–n diodes and bipolar transistors, with the models concentrating on the rate of change of stored charge. It is expected that the reader may require additional reading, for example references 8, 9, 30–33.

3.4.2 *p–n junctions*

To simplify the discussion, let us consider the p^+–n diode of Fig. 3.12, where the p^+-contact has two or three orders of magnitude more impurity doping than the n-region.

On reverse bias, the positive voltage on the n-material and the negative voltage on the p-material attract electrons and holes, respectively, pulling

Fig. 3.12. Schematic diagram for charges in a p^+–n junction. (*a*) Reverse bias - electrons pulled away from p^+-region by reverse bias voltage $+V_r$ to *n*-contact leaving depletion region with $d \propto V_r^{1/2}$. (*b*) Forward bias. Contact to p^+-region positive with respect to n-contact, which can pass only electrons. Charge Q (holes) stored in n-side until recombination turns hole current (holes travelling to the right) into electron current (electrons travelling to the left). Recombination time $= \tau_r$.

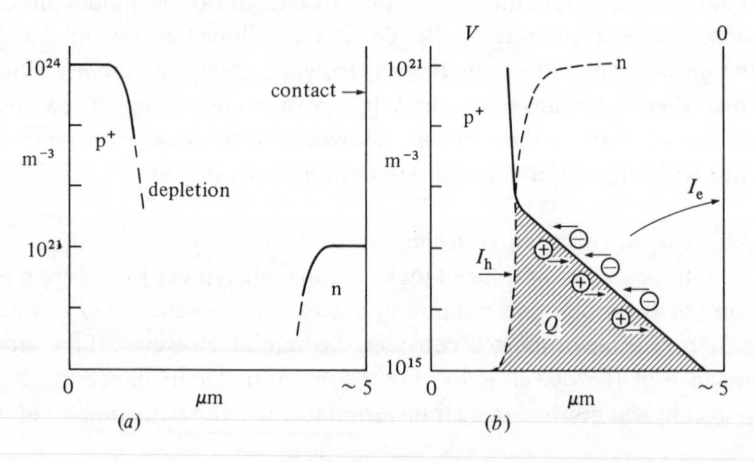

the charge carriers apart. The only source of charge carriers are thermally generated electrons in the p-side, which can travel to the positively biased n-contact, and similarly the thermally generated holes in the n-side, which can travel to the negatively biased p-contact. However, thermal generation has to give sufficient energy to the valence electrons for them to overcome the binding energy (which in materials like Si and GaAs is over 1 eV, while the average thermal energy is only the order of 25 meV at room temperature). It is found then that such thermally generated currents can lie in the picoamp range and so often be neglected.

On forward bias, the main source of carriers is the p^+-contact giving holes which diffuse over the Fermi potential barrier which lies between the n- and p-materials. The holes are attracted by the field to the negatively biased n-region. However, the holes cannot escape into the external circuit because the physics of this n-type contact is such that it can interact with and transmit only conduction electrons. (Similar remarks apply for electrons flowing into the p-side but, given low enough doping of the n-source, these contribute much less current.) At first sight then current cannot flow and a build up of charge Q of excess minority carriers (holes) must occur in the n-material. The conduction electrons can rapidly rearrange to neutralise any excess positive charge at the rate determined by dielectric relaxation, but with a neutral material there is no field to move the holes so that the excess numbers of holes can be removed only through recombination (Fig. 3.12). The rates of recombination must determine the current flow in the diode.

If a charge Q of holes is stored in a n-region and the holes recombine with the electrons at a rate Q/τ_r then there must be a current $-I$ of electrons from the n-contact matching the same current I of holes diffusing in. If the charge Q increases then there must be an additional current dQ/dt. The total convection current contributed by the hole-electron recombination is therefore

$$I = (dQ/dt) + (Q/\tau_r) \tag{3.4.1}$$

Recombination time is in general so much longer than the transit time that by comparison the transit time of the charge across the diode hardly affects the switching rate and is normally not considered.

The charge Q has to surmount a potential barrier, and in equilibrium a charge Q_0 manages to diffuse from the p-region into the n-region, but is balanced by an equal and opposite flow of thermally generated holes to give no net hole current. On forward bias the potential barrier is lowered by an amount V so that from classical Boltzmann statistics one expects a charge, $Q_0 \exp(eV/kT)$ to now surmount the potential barrier, while the opposing thermal generation is unaltered. On this basis the stored charge (neglecting transit time effects) is

$$Q = Q_0[\exp(eV/kT) - 1] \tag{3.4.2}$$

(3.4.1) and (3.4.2) then give a dynamical model of the p–n diode, taking into account the first-order rates of removal of charge by recombination. The addition of electron flow from the n- to p-side simply adds more charge recombining at a slightly different rate so that two time constants may appear in nearly equally doped p–n diodes.

In the steady state

$$I = I_0[\exp(eV/kT) - 1] \tag{3.4.3}$$

where

$$I_0 = Q_0/\tau_r$$

The region between the p- and n-contacts forms a capacitance C_d, and typically this capacitance changes as the bias voltage is altered. For a uniformly doped diode with abrupt junctions between the p- and n-material it is found that

$$C_d = C_{d0}(1 - V/V_f)^{-1/2} \qquad (V < V_f) \tag{3.4.4}$$

where V_f is the built in Fermi potential barrier between the p- and n-material and C_{d0} is the capacitance at zero bias. Capacitive current must be added to the convective current so that the total current for the diode should become

$$I_d = (\mathrm{d}Q/\mathrm{d}t) + (Q/\tau_r) + \mathrm{d}(C_d V)/\mathrm{d}t \tag{3.4.5}$$

On reverse bias with V negative, the diode capacitance C_d reduces as the holes and electrons are pulled apart by the bias and the depletion region widens. Diodes can be designed so that the depletion region runs into an n^+-type contact which limits further widening of the depletion region and so leads to a nearly constant reverse bias capacitance. The total charge storage effect is then highly non-linear with voltage and such diodes can be used as mixers and generators of harmonics of microwave frequencies[18] (see problem 3.5).

Earlier, p–i–n diodes were mentioned with an i-region reducing the capacitance on reverse bias. Such p–i–n diodes are still approximately modelled by (3.4.5). On forward bias, the i-region is flooded with a plasma formed by holes (charge Q) and neutralising electrons which are recombining (recombination time τ_r). The time taken to switch the plasma on and off is then determined primarily by this recombination time. However, once the plasma is formed then the diode can conduct high frequencies well, like any plasma switch, and the p–i–n diode is used extensively as a microwave switch.[18]

3.4.3 *The bipolar transistor*

To be definite we consider a p^{++}–n^+–p–p^+ transistor where the p^{++}-emitter region is doped two or three orders more than the n^+-base region, which in turn is much more highly doped than the p-collector region (Fig. 3.13). The final p^+-region is a good contact to the p-collector. The simple charge storage model highlights again the rates at which charge can be transported as one of the factors determining the switching speeds.

In normal active operation, the base–emitter junction is forward biased so that, with the highly doped emitter, the main current is formed from holes driven from the emitter into the base. These holes diffuse over the potential barrier between the p^{++}- and n^+-materials, exactly as in a diode. As in the diode, the contact to the n^+-base can transmit only conduction electrons. A charge Q of minority holes is then stored in the n-type base. As for the diode, this charge is controlled by the Boltzmann law:

$$Q = Q_0[\exp(eV_{be}/kT) - 1] \tag{3.4.6}$$

The electrons can readily move in from the base contact to neutralise the charge of these excess minority holes, but the base contact cannot transmit the holes to the external circuit. The holes have to recombine with electrons or escape into the collector region. The base is thin, and may even have its impurity content designed to give a built in electric field to force the electrons over to the collector region, which is also biased negatively with respect to the base so that holes have to travel to the collector once they have passed through the base.

Fig. 3.13. Schematic cross-section for p–n–p bipolar transistor. Base contact to n-material cannot transmit holes. Holes in base have to recombine with electrons to give I_b, or diffuse into collector to give current I_c.

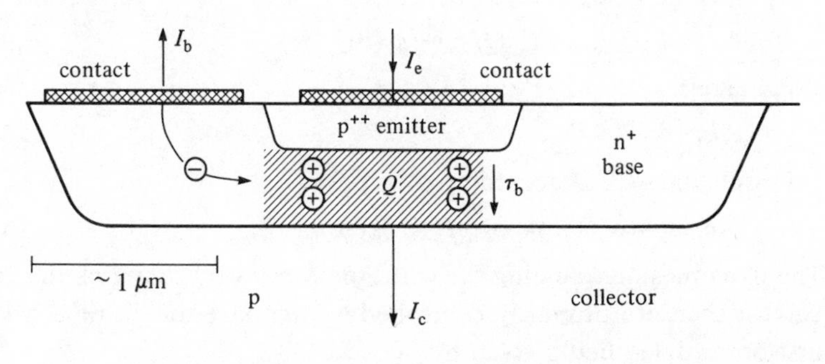

The scene then is set for elementary charge storage physics. If the transit time of the electrons across the base is τ_b ($\sim$10 ps/μm) then a principle current I_{cc}, flowing from the emitter to the collector and taking a time τ_b, has to store a charge of holes $Q = I_{cc}\tau_b$. These holes stored principally in the base region as minority carriers recombine with the electrons. The electron recombination current $I_b = Q/\tau_r$ comes from the base contact, but the corresponding hole currents have to come partly from an increase in the emitter current and partly from a decrease in the collector current so that

$$I_e = I_{cc} + I_{er} = (Q/\tau_b) + (K_1 Q/\tau_r) \tag{3.4.7}$$

$$I_c = I_{cc} - I_{cr} = (Q/\tau_b) - (K_2 Q/\tau_r) \tag{3.4.8}$$

where the total recombination currents

$$I_{er} + I_{cr} = I_b = Q/\tau_r \tag{3.4.9}$$

with

$$K_1 + K_2 = 1 \tag{3.4.10}$$

In general $\tau_b \ll \tau_r$, so that one may show using (3.4.10) that

$$I_c/I_e = \alpha_0 \approx \tau_r/(\tau_r + \tau_b) \tag{3.4.11}$$

This idealised value of α_0 is reduced in practical devices because not all the current is carried by the emitted holes. There would be some electrons which could not contribute to the collector current flowing from the base into the emitter. There could also be leakage currents at the edges of the base–emitter junction in the transistor.

When the dynamic operation is to be considered, the base charge changes by $\mathrm{d}Q/\mathrm{d}t$, and this requires additional current which again must come partly from an increased emitter current and reduced collector current. In certain problems the details of the division of the currents are not required and one may write

$$I_e = (Q/\tau_b) + (K_1 Q/\tau_r) + k_1(\mathrm{d}Q/\mathrm{d}t) \tag{3.4.12}$$

$$I_c = (Q/\tau_b) - (K_2 Q/\tau_r) - k_2(\mathrm{d}Q/\mathrm{d}t) \tag{3.4.13}$$

where again

$$k_1 + k_2 = 1 \tag{3.4.14}$$

For changes of Q according to $\exp j\omega t$

$$I_c/I_e \approx \alpha_0(1 - jk_2\omega\tau_b)/(1 + jk_1\omega\tau_b) \tag{3.4.15}$$

The dynamics of changing the collector current by changing the emitter current then are primarily controlled by the base transit time, with the first-order delay being given by

$$\alpha_b(\omega) \approx \alpha_0/(1 + j\omega\tau_b) \tag{3.4.16}$$

Turning to the control of the collector current by the base current

$$I_b = I_e - I_c = (Q/\tau_r) + (dQ/dt) \tag{3.4.17}$$

Again to a first order in the delay

$$\beta(\omega) \approx \beta_0/(1 + j\omega\tau_r) \tag{3.4.18}$$

where

$$\beta_0 \approx \tau_r/\tau_b$$

More detailed models are required to go much further.[1,2,31,33] The effects of the resistance r_b between the base contact and the active part of the emitter, along with the collector-base capacitance C_{bc}, provides an additional storage circuit when one wishes to change the base current. The maximum frequency for a transistor just with sufficient gain to oscillate can be shown to be

$$f_{max} \sim (2\pi/\tau_b r_b C_{bc})^{1/2} \tag{3.4.19}$$

Such considerations are left for further reading.[33]

From our simplified model, it can be seen that in 'common emitter' circuits where the base current controls the collector current[32] there is a current gain β. With a base current $i_b \cos \omega t$ the collector current has a magnitude $i_c = \beta_0 i_b/|1 + (\omega\tau_r)^2|^{1/2}$ so that high frequency responses, or equally fast switching speeds require short recombination times in the base of a transistor. But too short recombination times also reduce the current gain β_0.

Circuits in which the emitter current controls the collector current ('common base' circuits) always have a current gain $|\alpha(\omega)| < 1$ so at first sight appear less useful circuits. However, the collector current can be driven into a high enough impedance so that the output power is greater than the input power with the power gain falling off roughly as $|\alpha(\omega)|^2$. From (3.4.16) the frequency at which this power gain reduces by a factor of two is approximately β_0 times higher than the frequency at which the common emitter power gain is reduced by two. These circuits are potentially the most useful.

For example, one way of looking at the classic long tail pair circuit (Fig. 3.14) is that the input transistor acts like an emitter follower with the output transistor acting in a common base mode; both transistors are primarily limited in their performance by $\alpha(\omega)$. The 'cascode amplifier'[32] is another 'good' circuit. The input only requires a voltage gain close to unity and so has a good bandwidth, while the output circuit is again a common base transistor limited in its performance by $\alpha(\omega)$.

The model also demonstrates other design features. The importance of the base transit time in limiting the speed of the transistor suggests that transistors using charge carriers with the greatest mobility or highest carrier velocities will be advantageous. For Si and GaAs, n–p–n transistors will operate at high frequencies better than p–n–p devices.

One final point to make in comparing bipolar transistors with field effect transistors is that it is easier in the former to obtain higher rates of charge transfer than in the latter. The mutual conductance g_m can be defined for both the FET and the bipolar transistor. In the FET a $g_m \sim 20$ mS per 100 μm of gate length gives quite a good figure, while in the bipolar transistor the g_m defined by $\partial I_c/\partial V_{be} \approx I_0/(kT/e) \approx 40$ mS per mA of bias current I_0. The g_m of the bipolar depends not so much on the device structure as on the bias current level, and this generally leads to higher values of g_m than in the FET. A high g_m is often needed in computers where one device may have to drive several others.

PROBLEMS 3

3.1 In Fig. 3.1, for a fixed switch and source, show that the power in the load is a maximum if $R_L = R_i + R_s$. Show that the time constant is $(R_i + R_s)C_L/2$. Find the excess energy lost in the resistance R_s by C_L requiring additional charge per switching operation. Hence estimate the power lost for encoding at a bit rate B bit/s using such a switch.

3.2 Suppose that (Fig. 3.2) $\delta = t = W = 1$ μm and $A = 100\ \mathrm{W}^2$, with the capacitor formed from an oxide ($\varepsilon \sim 10^{-10}$ F/m). The conductor lead (length $L = 25$ μm) between the switching device and

Fig. 3.14. Fast switching circuits. (*a*) Long tail pair. (*b*) Cascode. T_2 in both circuits has grounded base (or common base) action. T_1 has low voltage gain and so responds rapidly to changes (high bandwidth).

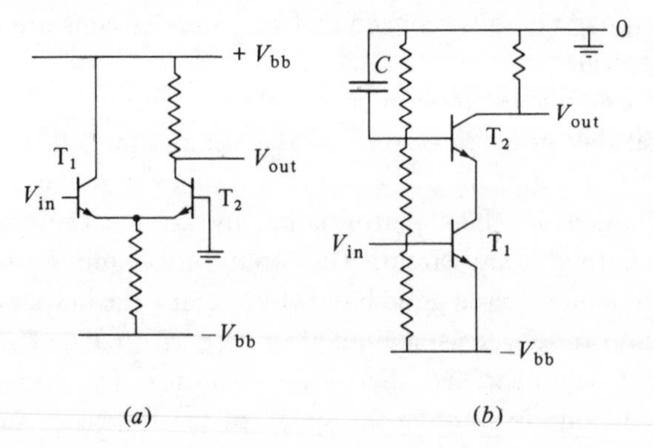

capacitive load is formed from a semiconductor with $10^{24}/\mathrm{m}^3$ donor impurities with a mobility of $0.01\ \mathrm{m^2/Vs}$. Estimate the limiting rise time.

3.3 On closing the switch in Fig. 3.P(a) show that the rate of change of charge on C is given from

$$LCQ'' + Q'[(L/R)+(CR)]+2Q = VC$$

Hence show that with $L=(3-2\sqrt{2})R^2C$ the characteristic switching time constant changes to $0.29RC$ from $0.5RC$ if $L=0$. Estimate the improvement in 10% to 90% rise time given by this 'maximally flat' L-value.

3.4 Suppose in a p–n diode with equal donor and acceptor concentrations N_i on either side, that the depletion depth on the p-side was $d+x$, while the depletion depth on the n-side was d. Show that x decreases to zero to give overall electrical neutrality on the same time scale as the dielectric relaxation time for the material. Assume equal mobilities μ for both p- and n-materials. With $n=10^{22}/\mathrm{m}^3$, $\varepsilon = 10^{-10}$ F/m, and $\mu = 0.1\ \mathrm{m^2/Vs}$, estimate at

Fig. 3.P. Inductive compensation - critical value can improve rise time and frequency response.

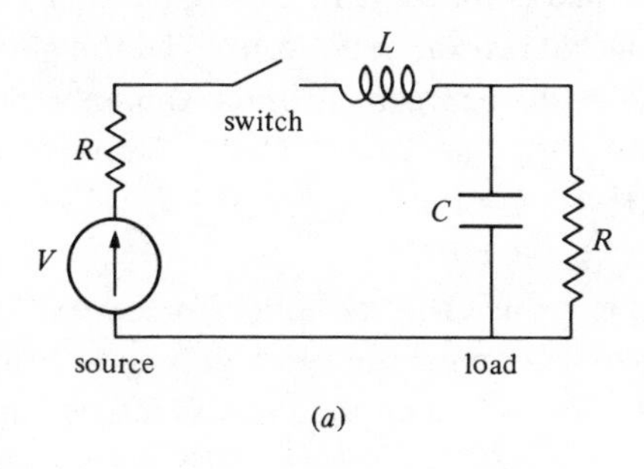

(a)

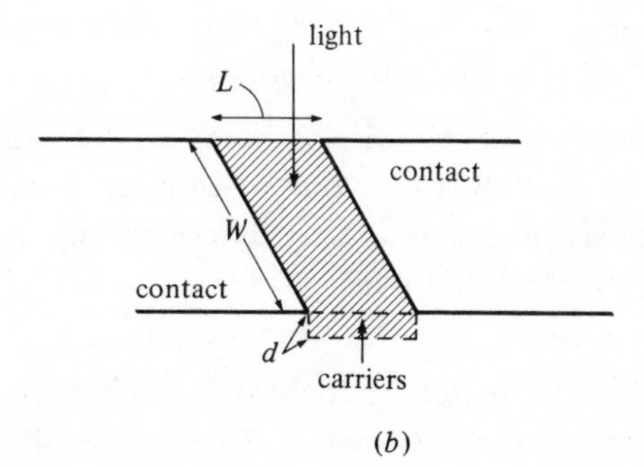

(b)

what frequencies one can no longer assume overall electrical neutrality for the p–n diode.

3.5 A 'step recovery' diode has a charge carrier lifetime of 200 ns. Estimate the charge stored Q_0 for a forward bias current of 10 mA. The voltage source V, fed through 50 Ω internal impedance, is changed to *reverse* bias the diode $V = 10[1 - \exp(-t/T)]$ V at time $t = 0$, where $T = 2$ ns. Estimate the time T_s taken to exhaust the initial stored charge Q_0. Take the voltage across the diode as negligible in comparison with 10 V until the stored charge is exhausted. The reverse bias capacitance is then to be 0.5 pF.

Estimate the time τ_{snap} after $t = T_s$ taken for the voltage to reach 9 V reverse bias (i.e. 90% of final value). Neglect transit time effects.

3.6 (*a*) A picosecond pulse of light has a total energy $\mathscr{E}$. The photon energy in this light is $\mathscr{E}_p = \hbar\omega > \mathscr{E}_g$. It is shone on a surface of a very high resistivity material, band gap $\mathscr{E}_g$. The material is antireflection coated, and the light is absorbed exponentially as $\exp(-x/d)$ with depth x, with η giving the quantum efficiency of hole–electron generation. The recombination time for the material is $\tau_r \gg 1$ ps. If the surface of the material is uniformly illuminated between two contacts (Fig. 3P (*b*)) L apart and W wide, show that the initial resistance R_s, caused by the photogenerated carriers between the contacts, is approximately independent of d (assumed $\ll L$) and of W ($\gg L$). Show that $R_s = L^2\mathscr{E}_p/[q\eta\mathscr{E}(\mu_n + \mu_p)]$. If $L = 1\ \mu$m, $\mathscr{E} = 1$ pJ, $\mathscr{E}_p = 1$ eV, $\mu_{n/p} = 0.05\ \mathrm{m^2/Vs}$, estimate R_s taking $\eta = 1$.

What factors might reduce the mobility from a bulk value?

(*b*) The light is now shone onto the same gap at a steady rate of R_{ph} photons/s. A voltage V is applied across the two contacts and the current changes by I_{ph} when the light shines steadily. Define a photoconductive gain $G = I_{ph}/qR_{ph}$. Show that $G = \eta\tau_r(\mu_n + \mu_p)\mathrm{V/L^2}$. Can G be larger than unity?

3.7 For a bipolar transistor, the minority carrier charge stored in the base is ρ and the recombination time within the base is τ_r, independent of position. Show directly by integrating the continuity equation (problem 2.7) that

$$I_e - I_c = (Q/\tau_r) + (\partial Q/\partial t)$$

independent of the distribution of the charge within the base, where Q is the excess minority carrier charge stored in the base.

Rapid dielectric relaxation keeps $E = 0$ in a uniformly doped base. Only diffusion current density $J = -D\, \partial\rho/\partial x$ flows (section 4.1). The minority carrier density ρ_0 at the emitter–base interface decreases approximately linearly over the base's width W to a low value (~ 0) at the base–collector interface where electric fields pull the charge rapidly away from the base at the limiting velocity. Estimate the charge stored in the base, the current flow across the base in terms of ρ_0, and hence show that an estimate for the transit time τ_b is $W^2/2D$. With $\mu = 0.1\ \mathrm{m}^2/\mathrm{Vs}$, $W = 0.3\ \mu\mathrm{m}$, $kT = 25\ \mathrm{meV}$, estimate τ_b. (Thinner bases or doping gradients in a base can decrease τ_b to the scattering limited values $\sim 10\,W$ ps, W in μm.)

3.8 Given cascaded delays

$$(\mathrm{d}V_1/\mathrm{d}t) + (V_1/\tau_1) = K_0 V_0(t); \quad (\mathrm{d}V_2/\mathrm{d}t) + (V_2/\tau_2) = K_1 V_1(t)$$

Show, to a first order, that the delay times add, i.e. $(\mathrm{d}V_2/\mathrm{d}t) + (V_2/\tau) = (\tau_1\tau_2/\tau) K_0 K_1 V_0(t)$, where $\tau = \tau_1 + \tau_2$. Show for an input $V_0(t) = A \exp j\omega t$, the error δV in this approximation is given from $|\delta V/V|^2 = (\omega^2\tau_1^2\tau_2^2)/[1 + (\omega\tau)^2]$. Adding delay rates is a useful approximation for $\omega < 1/\tau$.

4

Rates of change and transfer in phase space

4.1 The Boltzmann equation

4.1.1 *Introduction*

The motion of charge carriers within a semiconductor (or a gas) can be found by a more formal approach through the Boltzmann collision equation,[11,36–40] which provides an elegant method by considering the rates of change of particles within 'phase' space – a concept to be introduced shortly. To keep the discussion clear, a one-dimensional classical approach will be considered with charge carriers having an effective mass, m^*, assumed to be independent of energy or direction (not strictly valid in a semiconductor but still a most useful simplification). Extensions to three dimensions and the required corrections for quantum theory can be dealt with in later reading.

The Boltzmann collision equation for the flow of many particles is a statistical equation on the conservation of particles in a six-dimensional space referred to as phase space. The six dimensions consist of the three space dimensions for $\boldsymbol{x}$ combined with the three additional dimensions for momentum $\boldsymbol{p}$. Momentum is considered here as an independent variable with the same independence as position. It is the dynamical equations which link these six independent variables together. On first acquaintance, this concept of momentum and position being independent variables appears absurd because it is easy to confuse the dynamical link (given through an equation such as $m\,\mathrm{d}x/\mathrm{d}t = p$), with functional interdependence of p and x.

W. R. Hamilton in 1834 introduced the idea that all dynamical motion could be described in terms of a function $H(p, x, t) = 0$ linking the momentum p and position x in time. Coordinates of position and momentum can be defined and are treated as of equal independence. The dynamics then links the variables through their rates of change (Hamilton's equation and the hamiltonian energy function H):

$$dx_i/dt = \partial H/\partial p_i; \quad dp_i/dt = -\partial H/\partial x_i \tag{4.1.1}$$

with i running over all the separate coordinates.[41]

Related techniques are found in control theory where with state space variables n-dimensional second-order differential equations are reduced to $2n$ first-order differential equations. The $2n$ variables have an independent status.

Phase space can be conveniently demonstrated using the example of the harmonic oscillator, where the standard equation for motion of a particle of mass m^* with a restoring force kx is then

$$m^* \, d^2x/dt^2 = -kx \tag{4.1.2}$$

The momentum $p = m^* \, dx/dt$ is now considered to be an independent variable leading to the above equations

$$dp/dt = -kx; \quad dx/dt = p/m^* \tag{4.1.3}$$

(in accord with (4.1.1), where $H = (p^2/2m^*) + (kx^2/2)$). This one-dimensional motion is described with a solution in two-dimensional phase space:

$$x = A \sin(\omega t + \theta); \quad p = (A\omega/m^*) \cos(\omega t + \theta) \tag{4.1.4}$$

where $\omega^2 = k/m^*$. With appropriate scaling, the harmonic motion is a 'circle' in phase space (Fig. 4.1). If damping is included then the trajectory can spiral inwards (see problem 4.1). An external force will also modify the trajectory.

For a single particle, the two independent initial conditions (p_0 and x_0 at $t = 0$) determine uniquely the trajectory in phase space. For large numbers of particles, each particle has its own p–x–t trajectory so that there are many trajectories in the neighbourhood of any one point in

Fig. 4.1. Phase space trajectory for harmonic oscillator. Momentum and position scaled to give circle representation. Dashed line indicates schematically the effect of loss.

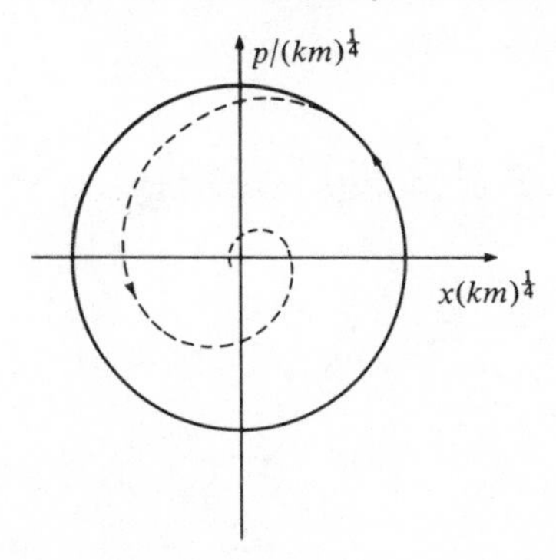

phase space. These trajectories cannot cross because if two particles have the same 'initial' conditions at any reference time then they are following the same path. Trajectories of neighbouring particles in phase space have closely similar trajectories, and one can then follow in a sensible manner the motion of a localised group of different particles, at least for small changes δp and δx (see Fig. 4.2).

4.1.2 *The distribution function*

The number of particles N_0 between x and $x+\mathrm{d}x$ and between p and $p+\mathrm{d}p$ is, for small enough $\mathrm{d}p$ and $\mathrm{d}x$, proportional to $(\mathrm{d}p\,\mathrm{d}x)$. The proportionality factor is $f(x, p, t)$, the probable density of particles in phase space. This function f is the *distribution function.*

$$N_0 = f(x, p, t)\,\mathrm{d}x\,\mathrm{d}p \tag{4.1.5}$$

The actual value of N_0 will have statistical fluctuations, but the fluctuations are ignored by the process of taking an average over a large enough *ensemble* of identical systems. All the systems in this ensemble are assumed to obey the same physics and have the same physical boundaries and same constituents, but the particles start with different initial conditions in each of the systems. Given M such 'identical' systems in this ensemble, then the expected numbers in the element $\mathrm{d}p\,\mathrm{d}x$ of phase space is $N = MN_0$. The fluctuations, being all statistically independent, are expected to be of the order of $N^{1/2}$ so that the fractional fluctuations decrease as $1/N^{1/2}$ giving the averages a sensible meaning over a large enough ensemble. The function f is then an ensemble average.

The density of particles in space must include all values of momentum so that

$$n(x, t) = \int f(x, p, t)\,\mathrm{d}p \tag{4.1.6}$$

The particle current density is

$$J_n(x, t) = \int (p/m^*) f(x, p, t)\,\mathrm{d}p = nv \tag{4.1.7}$$

Fig. 4.2. Trajectories of groups of particles in phase space.

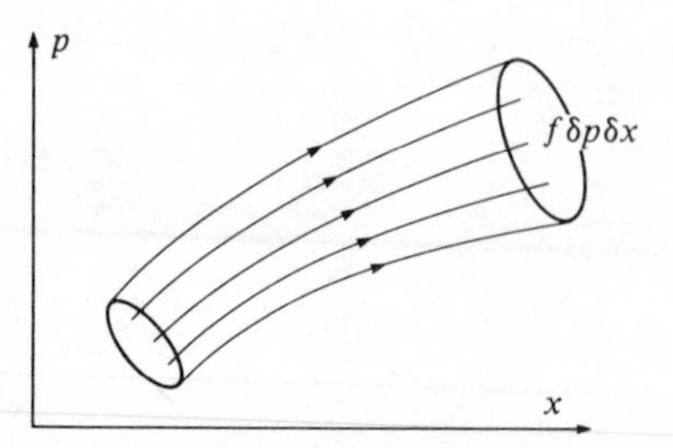

where the integration is again performed over all possible values of momenta and v is the average velocity of the particles.

For electrons the charge density is en and the electrical current density is $J = env$.

In thermal equilibrium it is known that there is a thermal energy of $\frac{1}{2}kT$ per degree of freedom of motion. So for an ideal gas in three dimensions (three degrees of freedom), the random kinetic energy gives the thermodynamic temperature

$$3kT/2 = \langle(\boldsymbol{p} - m^*\boldsymbol{v}) \cdot (\boldsymbol{p} - m^*\boldsymbol{v})/2m^*\rangle \tag{4.1.8}$$

where $\langle\ \rangle$ implies the average value.

For a 'gas' of conduction electrons in an electric field within a solid, (4.1.8) is an approximation because a thermodynamic temperature applies strictly to equilibrium. However, for charge carriers in semiconductors, the average velocity v is usually much smaller than the random velocity, and considerable simplification ensues by approximating further and defining an electron temperature T_e in terms of the total energy. As for (4.1.8), this electron temperature only represents a thermodynamic temperature for conditions where v is small. (Reference 38, for example, gives a more general discussion for a gas or fluid.) Consequently, define for one-dimensional flow

$$nkT_e = (1/m^*) \int [p^2 f(x, p, t)]\, \mathrm{d}p \tag{4.1.9}$$

Again the integral is understood to be over all possible momenta.

The general motion of the electrons within a semiconductor requires definition of the temperature T_e, the mean velocity v, and the carrier density n. All of these quantities can vary with position and time.

4.1.3 *Rates of change of the distribution function*

The distribution function f changes with time on account of several factors. The first simple reason can be that $f(x, p, t)$ is an explicit function of time. It is easy to become confused here because it might be argued that x and p are functions of time for a given particle. The distribution function does not consider a single particle, but the numbers of particles with momenta close to p and positions close to x at any one time t. If the bias currents and voltages applied to a device vary with time then the distribution of electrons within the device will vary with time. However, if the bias values are constant then the distribution of electrons would be expected to be constant with time, though generally varying in space.

The first direct rate of change to be considered is then

$$\mathrm{d}f/\mathrm{d}t)_{\text{time}} = \partial f/\partial t \tag{4.1.10}$$

Now consider a block of particles moving in phase space (Fig. 4.3). As the particles move a distance δx, so the distribution function has to change

$$\delta f = (\partial f/\partial x)\delta x \tag{4.1.11}$$

but

$$\delta x = (\mathrm{d}x/\mathrm{d}t)\delta t \tag{4.1.12}$$

Hence, the rate of change with time caused by the spatial motion is

$$\begin{aligned} \mathrm{d}f/\mathrm{d}t)_{\text{space}} &= \lim_{\delta t \to 0} \delta f/\delta t = (\partial f/\partial x)(\mathrm{d}x/\mathrm{d}t) \\ &= (\partial f/\partial x)(p/m^*) \end{aligned} \tag{4.1.13}$$

Similarly, since momentum space is on the same footing as conventional space, a movement of the batch of particles around p to $p+\delta p$ leads to a rate of change of f given by

$$\begin{aligned} \mathrm{d}f/\mathrm{d}t)_{\text{momentum}} &= (\partial f/\partial p)(\mathrm{d}p/\mathrm{d}t) \\ &= (\partial f/\partial p)F \end{aligned} \tag{4.1.14}$$

where $F = \mathrm{d}p/\mathrm{d}t$ is the force, the same for all values of momentum p. With the one-dimensional motion of electrons $F = eE$, though with three-dimensional motion we need to have $\boldsymbol{F} = e\boldsymbol{E} + e(\boldsymbol{p}/m^*)\times\boldsymbol{B}$. This extension is left as a problem (problem 4.2).

Adding all the rates of change together the total rate of change (or the *total* derivative) is

$$\mathrm{D}f/\mathrm{D}t = (\partial f/\partial t) + (p/m^*)(\partial f/\partial x) + F(\partial f/\partial p) \tag{4.1.15}$$

4.1.4 *Collisions*

Charge carriers moving through a crystal can collide or interact with each other. They can collide with defects or impurities in the crystal

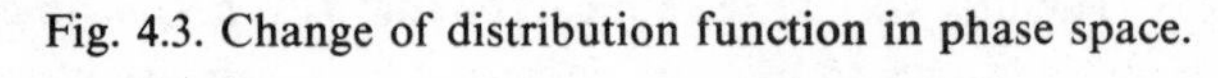

Fig. 4.3. Change of distribution function in phase space.

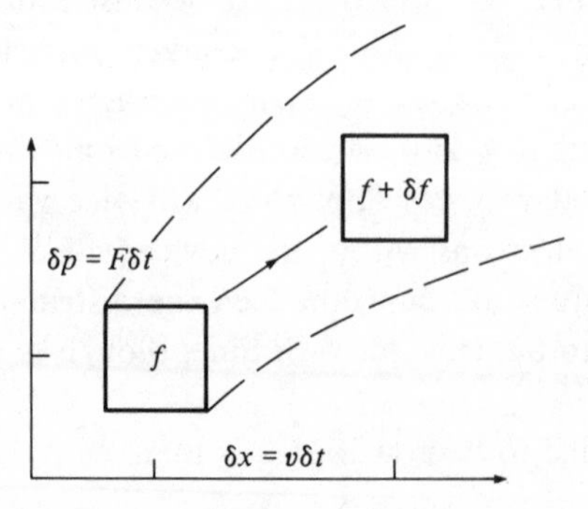

and, as already discussed, they can lose energy to the lattice vibrations (phonons) through the quantum and electromagnetic forces which hold the electrons and lattice together. It is convenient to lump all these effects together, at least initially, and assume that there is an interaction which occurs over a negligibly short time but leaves the particles free to move between the interactions or collisions. The collision then does not initially change the position of the particle but changes only its momentum, which can also lead to a concomitant change in energy if $|p|$ changes.

Although on having a collision any particular particle has its momentum changed in direction if not in magnitude, the distribution function $f(x, p, t)$ does not change its value of p. It is the value of f which changes because particles are scattered, by the collision, from a value p to some other momentum p_s. The rate of loss of particles of momentum p to other values of momentum must be proportional to the number of particles with the value p and so can be written as

$$\mathrm{d}f/\mathrm{d}t)_{\text{collisions out}} = -f(x, p, t)/\tau(p)_c \tag{4.1.16}$$

where $\tau(p)_c$ is the time scale for the scattering process.

While particles are moving out from the momentum p at (x, t) because of collisions, other particles are moving in. To make progress it is helpful to simplify this scattering of particles from other values of momentum back to the value p by assuming that these particles which are scattered back in have the effect of tending to restore an equilibrium distribution $f_o(p, x, t)$. Thus

$$\mathrm{d}f/\mathrm{d}t)_{\text{collisions in}} = f_o(x, p, t)/\tau(p)_c \tag{4.1.17}$$

leading to

$$\mathrm{d}f/\mathrm{d}t)_{\text{net}} = -(f - f_o)/\tau_c \tag{4.1.18}$$

The Boltzmann equation then balances the total rate of change of the distribution function $(\mathrm{D}f/\mathrm{D}t)$ with the rate of changes caused by the collisions so that

$$\mathrm{D}f/\mathrm{D}t = -(f - f_o)/\tau_c \tag{4.1.19}$$

The distribution function then tends to its equilibrium value f_o with a time scale of τ_c, possibly dependent on p. This time scale could be estimated by a more detailed theory or by experiment. For semiconductors considered here τ_c is typically subpicosecond.

4.1.5 *Equilibrium distribution*

In chapter 5, rate equation methods are given for finding equilibrium distributions. For the present, the equilibrium distribution is defined to be the Maxwell–Boltzmann distribution:

$$f_o = N(x, t) \exp(-\boldsymbol{p} \cdot \boldsymbol{p}/2mkT) \qquad (4.1.20)$$

(In one dimension the scalar product $\boldsymbol{p} \cdot \boldsymbol{p}$ is replaced by p^2). Although the Maxwell–Boltzmann distribution is a classical distribution, it is still valid for quantum particles, such as electrons, provided that the density of the electrons is not too great. The probability of an electron occupying any one energy state must be much less than unity, and also the energy states must be sufficiently closely spaced to say that there is approximately a continuous range of energies available to the electron.

Collisions which redistribute the particles in phase space cannot change the total numbers. (There is no consideration of loss of particles by processes such as recombination at this stage.) It follows that there has to be a restriction so that the total numbers which are scattered out have to be scattered back in, albeit with changed momenta:

$$\int (f_o/\tau_c)\, \mathrm{d}p = \int (f/\tau_c)\, \mathrm{d}p \qquad (4.1.21)$$

To simplify future discussion, it will be assumed that τ_c is independent of the momentum p. Then

$$\int (f_o)\, \mathrm{d}p = \int (f)\, \mathrm{d}p = n(x, t) \qquad (4.1.22)$$

4.1.6 *The continuity equation*

Writing out (4.1.19) in full gives

$$\partial f/\partial t + (p/m^*)(\partial f/\partial x) + F(\partial f/\partial p) = -(f - f_o)/\tau_c \qquad (4.1.23)$$

Now integrate both sides over all possible values of p using relationship (4.1.21), which eliminates the effect of collisions. The order of integration may be interchanged for x and t so that

$$\int (\partial f/\partial t)\, \mathrm{d}p = (\partial/\partial t)\left(\int f\, \mathrm{d}p\right) = \partial n/\partial t \qquad (4.1.24)$$

(using (4.1.6) for n). Similarly

$$(e/m^*)\int (p\partial f/\partial x)\, \mathrm{d}p = (e/m^*)(\partial/\partial x)\left(\int pf\, \mathrm{d}p\right)$$
$$= \partial J/\partial x \qquad (4.1.25)$$

(using (4.1.7) for J).

The force F given by an electric field is independent of p so that there is left an integral I given by

$$I = \int (\partial f/\partial p)\, \mathrm{d}p = f(p_{\max}) - f(-p_{\min}) \qquad (4.1.26)$$

However, by definition, $f(p_{max})=f(-p_{min})=0$ because the numbers of particles drop to zero for high enough magnitudes of momentum. Equivalently as $|p|\to\infty$, $f\to 0$ sufficiently fast that the integral I vanishes.

It will also be useful for later use to evaluate, using integration by parts:

$$I_u=\int [p^u\,\partial f/\partial p]\,\mathrm{d}p$$

the value

$$I_u=-u\int (p^{u-1}\,\partial f/\partial p)\,\mathrm{d}p \tag{4.1.27}$$

where u takes some integer value.

The result of I vanishing along with (4.1.24) and (4.1.25) yields then the continuity equation (section 1.2)

$$\partial J_n/\partial x+\partial n/\partial t=0 \tag{4.1.28}$$

4.1.7 *Momentum conservation*

The simple rate equation for momentum given in section 3.1 can now be augmented and justified better. Multiply (4.1.23) by p and integrate over all permissible values of p. As previously the integration with respect to p and differentiation with respect to x and t can be interchanged:

$$(\partial/\partial t)\left(\int pf\,\mathrm{d}p\right)+(\partial/\partial x)\left[\int p(p/m^*)f\,\mathrm{d}p\right]$$
$$+F\int p(\partial f/\partial p)\,\mathrm{d}p+\int p(f-f_o)(1/\tau_c)\,\mathrm{d}p=0 \tag{4.1.29}$$

Considering term by term using (4.1.7):

$$\int pf\,\mathrm{d}p=m^*nv \tag{4.1.30}$$

The equilibrium distribution has zero mean momentum

$$\int pf_o\,\mathrm{d}p=0 \tag{4.1.31}$$

Using integration by parts as in (4.1.27)

$$\int (p\partial f/\partial p)\,\mathrm{d}p=-n \tag{4.1.32}$$

From (4.1.9)

$$\int (p^2/m^*)f\,\mathrm{d}p=nkT_e \tag{4.1.33}$$

Combining all these results for (4.1.29) allows the rate of change of momentum to be arranged as

$$(1/n)\,\partial(nm^*v)/\partial t + (1/n)\,\partial(kT_e n)/\partial x + m^*v/\tau_c = F \qquad (4.1.34)$$

For many practical devices $(\tau_c/n)[\partial(nm^*v)/\partial t]$ is negligible compared to m^*v/τ_c so that it is customary to omit the first term in (4.1.34).

To complete this part of the discussion, define the mobility

$$\mu = q\tau_c/m^* \qquad (4.1.35)$$

and define a diffusion constant D by the 'Einstein relationship'

$$D = kT_e\mu/q \qquad (4.1.36)$$

Hence, for uniform T_e, one arrives at the required result for electrons:

$$v = -\mu E - (D/n)\,\partial n/\partial x \qquad (4.1.37)$$

with electron current density $J = evn = -qvn$.

A positive field gives a negative force on the electrons driving them in the $-x$-direction, hence the negative sign for the contribution $-\mu E$. The large spread of velocities means that some electrons are moving backwards and some moving forwards even though there is a mean velocity v. At a positive spatial gradient in the carrier density, there are more charge carriers moving back than are moving forward. There is then a change in the transport of momentum caused by the thermal spread of velocities combined with a spatial gradient. The diffusion coefficient D defines the magnitude of this effect, with the electron temperature T_e usually being taken to be the temperature of the crystal lattice, unless the electric field is large (see section 4.2).

The result for hole flow in a valence band follows *mutatis mutandis* (see problem 4.3).

4.1.8 *More about collisions*

The analysis of scattering with a fixed scattering time τ_c failed to distinguish the different types of collision: collisions amongst the electrons themselves (homocollisions) and collisions with other entities (heterocollisions) such as the lattice vibrations (phonons). Assigning a collision rate $1/\tau_s(p)$ to homocollisions where collisions do not lose overall numbers, momentum or energy:

$$\int (p^s f_0/\tau_s)\,\mathrm{d}p = \int (p^s f/\tau_s)\,\mathrm{d}p \qquad (4.1.38)$$

for $s = 0$, 1 or 2. The collision term then does not contribute directly to changes of density, momentum or energy of the electron gas. (In more detailed work, such collisions can change the distribution function so that there are some effects from such collisions which can show up in strongly non-equilibrium conditions.)

It is also possible to have collisions which destroy momentum but lose little energy. For such nearly elastic collisions one has the numbers conserved and energy nearly conserved:

$$\int (f_o/\tau_c)\,dp = \int (f/\tau_c)\,dp \tag{4.1.39}$$

$$\int [p^2(f-f_o)/\tau_c]\,dp \approx k(T-T_e)/\tau_e \tag{4.1.40}$$

where $1/\tau_e$ is significantly smaller than the values of $1/\tau_c$ because the energy is nearly conserved in the collisions. None of these collisions can affect the particle continuity but will affect in different ways the contribution to the momentum or, as seen later, the energy exchange.

There are also conditions in which the numbers of charge carriers are removed. For example, with recombination (section 2.3)

$$\begin{aligned}\partial n/\partial t)_{\text{recomb}} &= -\int (f_o/\tau_r)\,dp + \int (f/\tau_r)\,dp \\ &= -(n-n_o)/\tau_r \end{aligned} \tag{4.1.41}$$

where τ_r is the recombination time and $n_o = \int f_o\,dp$ is now the equilibrium carrier density to which the recombination attempts to restore the actual density. The continuity equation then has to be changed to become (in one dimension)

$$(\partial n/\partial t) + [\partial(nv)/\partial x] = -(n-n_o)/\tau_r \tag{4.1.42}$$

Fortunately the time scale of recombination is often slow in comparison with other collision time scales τ_c, so that recombination is ignored in the momentum balance. However, with very fast recombination (on a picosecond scale) one must expect the recombination to remove momentum and so reduce the mobility.

In the next section, two classes of electrons are considered in the conduction band with numbers n_1 and n_2. The continuity equation still holds for each class separately, but the scattering term has to allow for electrons to leave class 1 and be scattered back from class 2. The continuity relation for each class of particle has to allow for removal and arrival from other classes:

$$(\partial n_i/\partial t) + [\partial(n_i v)/\partial x] = \text{arrival rate}_{ji} - \text{removal rate}_{ij} \tag{4.1.43}$$

The collision term in the Boltzmann equation then needs careful consideration to effect an appropriate answer in the integration.

4.1.9 *Equilibrium in a potential*

The Boltzmann equation by accounting for the rates of change of the distribution function in phase space provides the basis for the flow of charge carriers within a semiconductor (or for the flow of fluids or gases). The equilibrium distribution function f_o has not actually been used in the work so far except to assert that its value is zero at high enough values of momentum.

As a final problem of this section, consider a semiconductor at equilibrium with $v=0$ within an electric field

$$E=-\partial V/\partial x \tag{4.1.44}$$

derived from a potential V. (4.1.37) can be integrated to give

$$n=n_o \exp(qV/kT_e) \tag{4.1.45}$$

where n_o is some constant. A positive potential, attractive to electrons, increases the density, while a negative potential (increasing the potential energy of the electrons) decreases their density. Comparison of (4.1.45) with the Maxwell–Boltzmann distribution function (4.1.20) suggests that the general result for the equilibrium distribution at total energy $\mathscr{E}$ is:

$$f_o(\mathscr{E}) \propto \exp(-\mathscr{E}/kT_e) \tag{4.1.46}$$

This result will be confirmed later.

4.2 **Energy transport and the transferred electron**

4.2.1 *Classical energy transport*

In (4.1.37) for the charge carrier's velocity, it is assumed that the diffusion constant is known, implying that the 'temperature' of the gas of charge carriers is known. It is frequently assumed for a semiconductor that this temperature is the temperature of the crystal lattice. At low electric fields this is an excellent approximation. Electrons gain energy from the electric field and lose energy to the crystal lattice. Unlike the high densities of mobile electrons in metals, the lower densities of electrons in a semiconductor contribute little to the conduction of heat. The thermal conduction is determined by lattice vibrations which gain thermal energy from interaction with the mobile electrons. These electrons excite lattice vibrations (sound or ultrasound) whose energy then propagates through the crystal to the heat sink at the contacts. The temperature of the lattice is determined mainly by how well the semiconductor is connected to a heat sink. Good heat sinks on practical semiconductor devices are usually essential.

With strong enough electric fields present, the carriers gain energy which cannot be transferred sufficiently rapidly to the crystal lattice. The electric field then heats up the electrons, which behave as a gas of 'hot electrons' of temperature T_e even though the lattice remains at a lower temperature T_o which is determined by heat conduction and any heat sink. The rates of supply and of loss of energy can again be made more evident by use of the Boltzmann equation.

To reduce complexity, the one-dimensional discussion is continued, with energy $\mathscr{E}$ and electron temperature T_e related as previously:

$$\mathscr{E} = kT_e = (1/m^*) \int [p^2 f(x, p, t)]\,\mathrm{d}p \tag{4.2.1}$$

Energy transport is found from the Boltzmann equation (4.1.23) by multiplying by p^2/m^* and integrating over all permissible values of p:

$$\begin{aligned}(\partial/\partial t)&\left\{\int [(1/m^*)p^2 f]\,\mathrm{d}p\right\} + (\partial/\partial x)\left\{\int [p/m^*][(1/m^*)p^2 f]\,\mathrm{d}p\right\}\\ &+\left\{\int [(1/m^*)p^2 F\,\partial f/\partial p]f]\,\mathrm{d}p\right\}\\ &= -\int [(1/m^*)p^2(f-f_o)/\tau_c]\,\mathrm{d}p \end{aligned}\tag{4.2.2}$$

Integration by parts, as in (4.1.27), evaluates the term with the gradient in momentum space to give

$$I_1 = \int [(1/m^*)(p^2)F\,\partial f/\partial p]f\,\mathrm{d}p = -2Fvn \tag{4.2.3}$$

The magnitude of I_1 gives the rate of supply of energy from the electric field and this supply must be balanced by the energy losses.

The scattering terms try to restore the energy $\mathscr{E}$ back to an equilibrium value $\mathscr{E}_o$, which would be given by the temperature of the lattice. It is assumed here that heat capacity of the semiconductor material prevents sudden local changes of $\mathscr{E}_o$ caused by local exchanges of energy between electrons and the lattice. On average, the energy or heat can be removed at a sufficient rate through thermal conduction of the lattice vibrations to maintain an almost constant value of $\mathscr{E}_o$ for the lattice equilibrium energy that is distinct from the electron energy $\mathscr{E}$. With this assumption it is not then necessary to insert yet another rate equation to determine $\mathscr{E}_o$!

Evaluating the scattering terms then gives an integral I_2:

$$\begin{aligned} I_2 &= \int [(1/m^*)p^2(f-f_o)/\tau_c]\,\mathrm{d}p\\ &= (\mathscr{E} - \mathscr{E}_o)n/\tau_e \end{aligned}\tag{4.2.4}$$

where τ_e has been used to replace τ_c after the integration to distinguish this from the characteristic time for momentum loss. As argued in section 4.1.8, rates of loss of momentum and energy can occur at quite different rates, especially if the 'collisions' are elastic.

It should be recorded here that one expects that the scattering rates will depend on the momentum p, a vector, rather than the energy $\mathscr{E}$, a scalar. An accurate integration of I_2, particularly when in three dimensions, cannot then lead precisely to the final result with a simple time constant which is the same for all problems. Nevertheless, use of the single time constant makes a useful engineering compromise to physical modelling.

The spatial gradient term gives rise to a third-order product for momentum and is evaluated approximately here as an integral I_3:

$$I_3 = (\partial/\partial x)\left\{\int (p/m^*)[(1/m^*)p^2 f]\,\mathrm{d}p\right\}$$

$$\approx \partial(v\mathscr{E}n)/\partial x \qquad (4.2.5)$$

This integral of p^3 would depend in detail on the shape of the distribution f in momentum space, but to evaluate it as approximately the same as $v\mathscr{E}$ is an assumption which permits progress at the modelling and is not too seriously in error when there are many randomising collisions.

The final term in $\partial/\partial t$ is evaluated directly in terms of $\mathscr{E}$ to give overall the approximate energy relationship

$$[\partial(\mathscr{E}n)/\partial t]+[\partial(v\mathscr{E}n)/\partial x]+[n(\mathscr{E}-\mathscr{E}_o)/\tau_e]=\alpha Fnv \qquad (4.2.6)$$

The factor α is 2 here in this one-dimensional problem.

The rate of supply of energy from the field is on the right-hand side and has to balance the rate of loss of energy on the left hand side and give any rate of increase of energy. If there is no supply and no scattering losses of the energy (i.e. all scattering was perfectly elastic) then one arrives at the continuity of energy as in (1.2.6), with the particle velocity v being the velocity at which the energy travels.

The results for three-dimensional analyses are more complex, and the reader is referred elsewhere[11,38] for more general results. However, for our purposes a key factor comes in balancing the rates of loss and supply of energy, and a useful approximation recognises that not all the energy given by the field can be given to the thermal energy in one direction. The energy which is gained from the field is spread randomly by collisions over the three degrees of freedom. Keeping $\mathscr{E} \to kT_e$ as for one dimension, one expects the factor α to become $\frac{2}{3}$ instead of 2 in (4.2.6).

As discussed in chapter 3, it is necessary for the energy transfer per collision to be less than $\hbar\omega_p$, the maximum energy which the electrons can lose to the lattice in one interaction. At high fields, scattering limited velocities can be obtained from this Boltzmann theory by making $(\mathscr{E} - \mathscr{E}_o)$ approach a limit of $\hbar\omega_p$.

4.2.2 *Two valley conduction – Gunn effect*

An exciting development in the physics of semiconductors and devices came in 1963 when Gunn, Hilsum, Ridley and Watkins[42–44] all contributed in various ways to the discovery of the transferred electron effect which led to 'Gunn' oscillations at microwave frequencies in GaAs. Although, currently, computer models work out expressions derived from quantum theory for scattering in momentum space, a useful understanding is obtained by a more classical discussion[11] of the rates at which momentum and energy can be transferred between electrons within a semiconductor.

Fig. 4.4 shows an energy-momentum diagram for electrons in equilibrium within a conduction band, where there are two main valleys for the electrons – (i) highly mobile electrons with a light effective mass or (ii) relatively immobile heavier electrons. Because of the complex quantum wave functions that can exist for electrons within a three-dimensional periodic potential, as in a crystal, all semiconductors have several valleys in which electrons can lie with different effective masses and mobilities. At low fields this is not important because the electrons only congregate around the lowest energies in the conduction band.

Now if electrons gain energy from an electrical field E faster than they lose energy to the lattice, they may gain sufficient net energy to be able to transfer into the higher energy states. If these higher energy states have a low enough mobility then under certain conditions it is possible to find that the electrons' velocity v slows down to such an extent that

Fig. 4.4. Energy-momentum in two valley semiconductor. N_{ci} gives density of states, n_i = density of electrons occupying state i, μ_i gives the mobility $(i = 1, 2)$.

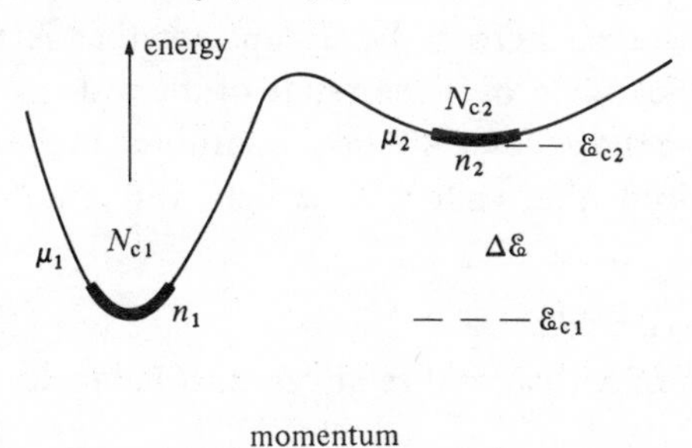

dv/dE becomes negative. This leads to a specimen of n-type GaAs material having a negative electrical resistance when suitably biased. Energy is given out by this negative resistance and, in appropriate circuits, oscillations occur in contrast to the damping of any oscillations with a normal positive resistance.

The two valley model indicated in Fig. 4.4 is adequate for discussions here, with n_i electrons in the ith valley ($i = 1, 2$), with effective mass m_i^*, mobility μ_i and effective densities of states N_{ci} which are known[29,35] to be proportional to $m_i^{*3/2}$. The lowest energy state of each valley is $\mathscr{E}_{c1}$ and $\mathscr{E}_{c2}$ such that $\mathscr{E}_{c2} - \mathscr{E}_{c1} = \Delta$, and consequently in equilibrium, according to standard Boltzmann statistics (section 5.1), it is suggested that

$$n_i = N_i \exp[-(\mathscr{E}_{ci} - \mathscr{E}_f)/kT_e] \tag{4.2.7}$$

or

$$n_2/n_1 = R \exp(-\Delta/kT_e) \tag{4.2.8}$$

where

$$R = N_2/N_1 = (m_2^*/m_1^*)^{3/2}$$

Normally, unless R is large enough and T_e is large enough, $n_2 \ll n_1$ and there is no effect caused by the very few electrons in the higher energy states. To explore this result further there are three important sets of rate equations to be examined – numbers, momentum and energy.

4.2.3 *Number conservation*

With a density of n_1 particles of a single species, it was shown in (4.1.43) how the one-dimensional continuity equation gives

$$[\partial n_i/\partial t] + [\partial(n_i v)/\partial x] = \text{arrival rate}_{ji} - \text{removal rate}_{ij} \tag{4.2.9}$$

The arrival rate from valley 2 will be proportional to n_2, the density of electrons available to transfer down to lower energies. In the model it is supposed that there are plenty of empty states in valley 1 so that lack of empty states does not restrict the numbers transferring down:

$$\text{arrival rate})_{\text{from 2 to 1}} = n_2/\tau_{21} \tag{4.2.10}$$

The time constant τ_{21} is taken here to be a constant that is roughly independent of the electrical field over the range of interest.

The rate of removal from the valley 1 will be proportional to the density available for removal into the valley 2, which will similarly be proportional to n_1:

$$\text{removal rate})_{\text{from 1 to 2}} = n_1/\tau_{12} \tag{4.2.11}$$

In equilibrium the rate of arrival and removal must balance:

$$n_{1e}/\tau_{12} = n_{2e}/\tau_{21} \tag{4.2.12}$$

Hence, on using (4.2.9) and (4.2.14)

$$1/\tau_{12} = (1/\tau_{21})R \exp(-\Delta/kT_{e1}) \qquad (4.2.13)$$

This result can be interpreted that the rate of scattering of *available* electrons from the lower energy to the higher energy is the same as the rate of scattering of available electrons from the higher to the lower energy. However, while all high energy electrons are available to transfer down in energy, only a fraction $R\exp(-\Delta/kT_{e1})$ of the n_1 lower valley electrons are available for transfer over the energy barrier Δ. It will be assumed that (4.2.13) approximately holds for all conditions with the continuity equations

$$(\partial n_1/\partial t) + [\partial(n_1 v)/\partial x] = (n_2/\tau_{12}) - (n_1/\tau_{21}) \qquad (4.2.14)$$

$$(\partial n_2/\partial t) + [\partial(n_2 v)/\partial x] = (n_1/\tau_{21}) - (n_2/\tau_{12}) \qquad (4.2.15)$$

4.2.4 *Momentum and energy balance*

The momentum balance for a single particle species now has to be modified. For the lower valley electrons, the field has to supply momentum $m_1^* v_1$ to those electrons which are scattered in from valley 2. The additional rate of supply of momentum required from the field is then $(n_2/\tau_{21})(m_1^* v_1)$, where it is assumed that these electrons arrive in valley 1 with zero mean momentum, leading to

$$n_1 F = [\partial(n_1 m_1^* v_1)/\partial t] + [\partial(\mathscr{E}_1 n_1)/\partial x] + (n_1 m_1^* v_1/\tau_{m1}) + (n_2 m_1^* v_1/\tau_{21}) \qquad (4.2.16)$$

Similarly for the upper valley

$$n_2 F = [\partial(n_2 m_2^* v_2)/\partial t] + [\partial(\mathscr{E}_2 n_2)/\partial x] + (n_2 m_2^* v_2/\tau_{m2}) + (n_2 m_2^* v_2/\tau_{12}) \qquad (4.2.17)$$

For the energy balance, consider the modifications required to (4.2.8). First subscripts must be added to indicate which valley is being considered. The energy $\mathscr{E}_1$ will be assumed to give the electron temperature T_{e1}, while the energy in the upper valley will be assumed to remain close to thermal equilibrium because it contains electrons which move sufficiently slowly that they gain energy from the field at too low a rate to become significantly hotter than the lattice. So

$$\mathscr{E}_1 = kT_{e1}; \quad \mathscr{E}_2 = kT_{e2} = kT_o \qquad (4.2.18)$$

Electrons scattering into the upper valley are mainly those electrons which have reached an energy Δ higher than the average and so take an energy $(\mathscr{E}_1 + \Delta)$ with them. Equally those electrons which scatter from valley 2 into valley 1 already have a potential energy Δ which can be converted into kinetic energy on transfer. Thus again balancing the rates of supply and loss, including these extra scattering mechanisms, modifies (4.2.6) to become in one dimension

$$[\partial(\mathscr{E}_1 n_1)/\partial t]+[\partial(v_1\mathscr{E}_1 n_1)/\partial x]+[n_1(\mathscr{E}_1-\mathscr{E}_o)/\tau_{e1}]$$
$$+[n_1(\Delta+\mathscr{E}_1)/\tau_{12}]-[n_2(\Delta+\mathscr{E}_2)/\tau_{21}]=\alpha eE(n_1 v_1) \quad (4.2.19)$$

As for the previous discussion, an appropriate value for α is $\frac{2}{3}$. A similar equation holds for valley 2, but it will not be needed because of the assumption just discussed that $T_{e2}\sim T_o$. The result of (4.2.13) is still assumed valid in the dynamic case as only a proportion of the hot electrons can transfer.

Every rate equation must give sensible results at equilibrium, and it is readily checked that in equilibrium with $\mathscr{E}_1=\mathscr{E}_2$ the energy balances with (4.2.12) holding.

4.2.5 *Velocity/field characteristics*

Applying the theory to GaAs, appropriate parameters must be obtained from experiments, parameters fitting to experiments, or more detailed theory. The effective mass of normal conduction band electrons in GaAs is known to be $m_1^*=0.067\ m_0$, and good material can have a mobility up to 0.85 m^2/Vs, giving $\tau_{m1}=0.32$ ps. The upper valleys are known to lie among the 8 equivalent ⟨111⟩ directions of crystal momentum. The minimum energy of these valleys is currently estimated to be 0.31 eV ($\approx 12.4\ kT_o$) above the minimum of the central valley. The ratio of the effective masses in the two types of valley is $m_2^*/m_1^*=0.55/0.67=$ 8.2. The effective density of energy states N_{ci}, which gives a measure of how many electrons can readily pack into the different valleys, is given from semiconductor theory as proportional to $m_i^{*(3/2)}$. In counting quantum states for electrons in a semiconductor, only states in the first Brillouin zone of the energy-momentum space must be counted. The upper valleys have their central minima at a value of crystal momentum which makes the valleys lie approximately half in the second Brillouin zone. The ratio of the effective density of states in 2 and 1 is then

$$R=\tfrac{1}{2}\times 8\times(m_2^*/m_1^*)^{3/2}=94$$

and the symmetry in k space permits one to treat the valleys as all being equivalent.

It is convenient to normalise the variables with the equilibrium thermal energy $\mathscr{E}_o$:

$$\mathscr{E}_o=0.025\ \text{eV}=4\times 10^{-21}\ \text{J}=m_1^* v_t^2 \quad \text{where } v_t=2.5\times 10^5\ \text{m/s}$$

$$E_t=v_t/\mu_1=3.2\times 10^5\ \text{V/m} \quad \text{where } \mu_1=e\tau_{m1}/m_1^*=0.8\ \text{m}^2/\text{Vs}$$

$$\tau_{m1}=3\times 10^{-13}\ \text{s} \quad \text{where } x_t=v_t\tau_{m1}=0.08\ \mu\text{m}$$

taking the nearest round figures. Time is then measured in units of τ_{m1}, energy in units of $\mathscr{E}_o$, velocity in units of v_t and field in units of E_t.

The normalised equations will be rewritten with all space variations ignored, but time variations retained to illustrate the dynamics of electron transfer which is proving to be so interesting in research with devices of small dimensions. The equations may be rearranged as

$$(\partial n_1/\partial t)+a_i n_1(S+1)=a_i \tag{4.2.20}$$

with $S=R\exp(-\Delta/\mathscr{E})$ and $a_i=\tau_{m1}/\tau_{21}$. The subscript 1 is dropped on the energy $\mathscr{E}$, and in normalised terms $n_1+n_2=1$.

The velocity and energy of electrons in valley 1 is given by

$$(\partial v_1/\partial t)+v_1\{1-a_i[(2+S)-(1/n_1)]=F \tag{4.2.21}$$

$$(\partial\mathscr{E}/\partial t)+(\mathscr{E}-1)\{a_e-a_i[1-(1/n_1)]\}$$
$$+\Delta a_i[(S+1)-(1/n_1)]=\alpha v_1 F \tag{4.2.22}$$

where the normalised force is $F=E/E_t$; $a_i=\tau_{m1}/\tau_{21}$; and $a_e=\tau_{m1}/\tau_e$.

The electrons in the upper valley have such a low mobility in this model that fast momentum relaxation forces v_2 to the equilibrium velocity much faster than other time scales, so approximately

$$v_2=F/[r(a_2+a_1)] \tag{4.2.23}$$

where $r=m_2^*/m_1^*$ and $a_2=\tau_{m1}/\tau_{m2}$.

In equilibrium, the net normalised velocity is

$$v=n_1v_1+(1-n_1)v_2 \tag{4.2.24}$$

with

$$v_1=F/(1+a_iS);\quad F^2=(3/2)(1+a_iS)(a_e+a_iS).$$

Fig. 4.5 shows the results evaluating the steady state velocity/field characteristics with $d/dt=0$. A useful fit with experimental results is found for $\tau_{e1}=0.5$ ps, and $\tau_{21}=1$ ps.

If the electric field remains uniform inside the sample of material then the electric field E is directly linked to the voltage V ($=E/$length), while the velocity v is directly linked to the current I ($=eNv\times$area). Ideal waveforms might be square waves which change from threshold voltage $V_t=E_tL$ and current I_t to a high voltage and low current giving out power at about 30% efficiency of conversion from bias power to microwave power (see problem 4.5). Unfortunately in practical Gunn devices it is not possible to achieve such ideal square waveforms and it is also difficult to make the field uniform. Efficiencies around 10% at 10 GHz are more usual. The non-uniformity of the electric field arises because the negative differential mobility makes the pattern of uniform charge unstable, leading to the spontaneous break up of the electric field into

Fig. 4.5. Velocity field characteristic for two valley, two temperature model. (*a*) Velocity field. (*b*) Energy in 1 and 2 per electron. (*c*) Density in 1. (*d*) Velocity 1 and 2 against field. Valley numbers = $\frac{8}{2}$; $m_2^*/m_1^* = 8.2$; $\tau_{m1}/\tau_{e1} = 0.6$; $\tau_{m1}/\tau_{12} = 0.3$; $\tau_{m1}/\tau_{m2} = 10$. Peak velocity $\sim 2 \times 10^5$ m/s at $E_t \sim 3.2 \times 10^5$ V/m.

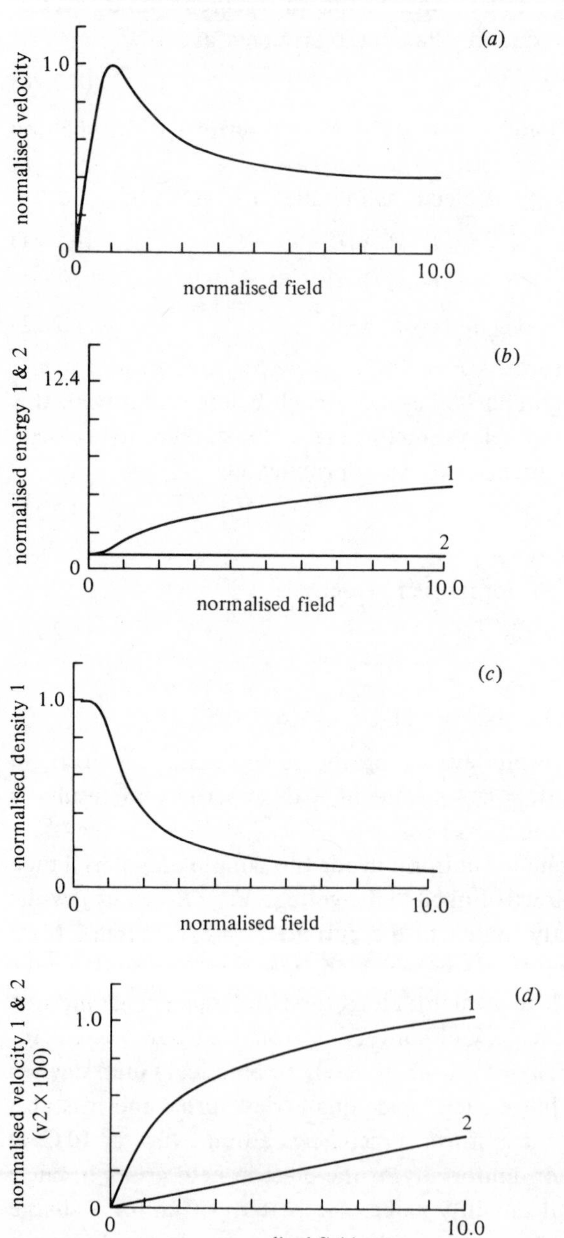

regions or *domains* of high field, a result first predicted by Watkins and Ridley[44] (see section 2.2 and problem 4.4.)

In early Gunn devices, operating at 1 GHz, domains easily had over 100 ps to form but yet be present for most of the cycle. This domain instability was of exceptional interest and forced oscillations with periods related to the time taken for domains to travel through the material. Indeed it is this difficulty in maintaining a uniform field which prevents one making 100 MHz Gunn oscillators. However, at 10 GHz, domains can take a significant fraction of a cycle to form and indeed can be inhibited from forming by switching the field fast through the values where $dv/dE < 0$. Domains can also be inhibited from forming in thin layers of material as in an FET. All this has reduced the interest in domain formation and domain characteristics. The emphasis in Gunn devices is more on the practical magnitude of the negative resistance that can be obtained along with the design of efficient microwave circuits.

Of importance, at high frequencies, is the time it takes for electrons to transfer from the lower valley with a high mobility to the upper valley with a low mobility. If electrons move suddenly into a high electric field then they switch over to the low mobility states in time, but within that time they accelerate up to velocities which are much higher than the scattering limited equilibrium value.[45,46] Fig. 4.6 demonstrates this by supposing that a group of electrons are suddenly injected into a field which is 5 × threshold from a lower value of about $\frac{1}{2}$ × threshold. The unit normalised time markers can be compared with the unit normalised distance markers, and it may be seen that the first 5 units of distance are

Fig. 4.6. Qualitative ballistic effects in intervalley transfer. Electron injected into field five times threshold from field of one-half times threshold. Time markers give about $\frac{3}{10}$ ps while total distance travelled is about 0.8 μm. Notice that in ballistic regime electrons accelerate, travelling much faster than normal. (Transfer parameters as for Fig. 4.5.)

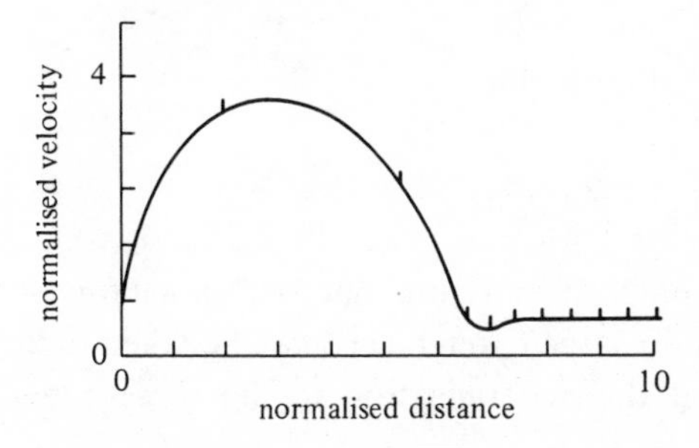

traversed in about 2 time units, while at the end the velocity is below 1 unit of distance for 1 of time. Thus on a distance scale of 0.1 μm the behaviour of the electrons is very different from equilibrium and the electrons are said to be moving in a 'ballistic' regime. This has important implications for FETs, where the electric fields are high and the gate lengths are submicron. The transit times are expected to be shorter than the simple velocity/field results would indicate, and an enhancement is found for high frequency FETs. Materials and device geometries are being examined where effectively there are longer net scattering times so that there is less energy lost.[29,35] Such systems would give even better results in the ballistic regime. An appreciation of the rates at which changes can occur gives the device designer and the materials scientist new goals and impetus to reach these goals of higher speeds and smaller dimensions.

PROBLEMS 4

4.1 Consider a harmonic oscillator with a small damping term so that (4.1.3) becomes $dp/dt = -kx - 2\alpha p$. Scale so that $P = p/(km)^{1/4}$, $X = x(km)^{1/4}$, $T = t(k/m)^{1/2}$. By writing $R = X + jP$ show that approximately

$$dR/dT = -jR - \alpha(m/k)^{1/2}R$$

and interpret this sketch using the Argand diagram as a plot in (X, P) phase space.

4.2 In (4.1.14), the three-dimensional generalisation leads to

$$df/dt)_{\text{momentum}} = \boldsymbol{F} \cdot \boldsymbol{\nabla}_p f,$$

where

$$\boldsymbol{\nabla}_p f = (\partial f/\partial p_1, \partial f/\partial p_2, \partial f/\partial p_3)$$

and

$$\boldsymbol{F} = e\boldsymbol{E} + e(\boldsymbol{p}/m^*) \times \boldsymbol{B}.$$

Evaluate

$$I_1 = \iiint \boldsymbol{F} \cdot \boldsymbol{\nabla}_p f \, dp_1 \, dp_2 \, dp_3$$

$$I_2 = \iiint p_1 \boldsymbol{F} \cdot \boldsymbol{\nabla}_p f \, dp_1 \, dp_2 \, dp_3$$

(Hint: integrate by parts and note that $\partial F_i/\partial p_i = 0$ for $i = 1, 2$ or 3; also note that for large enough $|p|$ f can be taken to be zero.) Hence show that the generalisation to the three-dimensional form (4.1.28) is

$\partial n/\partial t + \text{div}\, \boldsymbol{J}_n = 0$

and (4.1.37) becomes

$\boldsymbol{J} \approx qn\mu \boldsymbol{F} + qD \text{ grad } n$

4.3 Follow through the Boltzmann equation for holes and show that the current density is given by

$J_\rho = \mu\rho E - D\, \partial\rho/\partial x$

where ρ is the charge density of holes (ρ not to be confused with momentum p). The mobilities μ and diffusion constant D are naturally appropriate values for holes, different from the electrons' values.

4.4 High field 'domains' can spontaneously form inside a Gunn diode. Consider a small disturbance to the electron density n as in Fig. 4.Pa. Assume this disturbance travels at a constant velocity and that in the region of this velocity $\mathrm{d}v/\mathrm{d}E = -\mu_\mathrm{n}$, where $v(E)$ gives the velocity/field characteristic. Neglect diffusion. Using section 2.2 estimate the time scale for the growth from (a) to approximately (b), in terms of the negative dielectric relaxation time τ_dn. Noting that the background neutralising charge of ionised donors n_o is uniform, sketch the field E associated with this domain ($\varepsilon\, \mathrm{d}E/\mathrm{d}x = -q(n - n_\mathrm{o})$: Gauss).

4.5 Assuming that no domains form, $V = EL$ and $I = qnv(E)A$ for a transferred electron device (Gunn diode) of length L, cross-sectional area A and uniform electron density n. Consider a square wave of voltage from V_t to KV_t ($K > 1$) with I changing in the Gunn diode accordingly from I_t to $kI_\mathrm{t} = qnv(KV_\mathrm{t}/L)A$, ($k < 1$). Find the efficiency $\eta = P_\mathrm{rf}/P_\mathrm{dc}$, where P_rf is the power out at the fundamental frequency of the square wave and P_dc is the direct power in.

Fig. 4.P. 'Domain' formation.

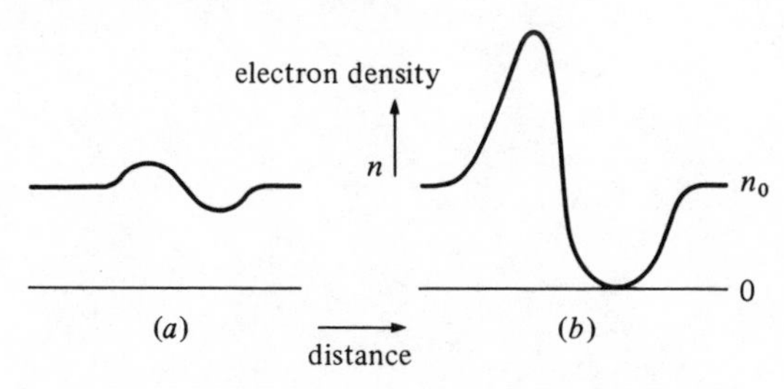

Experimentally this estimate for η is found to be a factor of two or more too high. Suggest reasons why.

4.6 Starting with

$J=\rho v(E)-D\,\partial\rho/\partial x, \qquad (\partial J/\partial x)+(\partial\rho/\partial t)=0,$

$\varepsilon\,\partial E/\partial x=\rho-\rho_0.$

Given mean values v_0, ρ_0, assume small changes ρ_1, J_1, E_1 etc. varying spatially as $\exp(-j\beta x)$. Neglect squares $E_1\rho_1$ etc. Note $v_1=(dv/dE)E_1$. Find an expression for the rate of growth of ρ_1 when $dv/dE=-\mu_n$.

Explain why such domains will not be expected to grow in the thin channel of a GaAs FET under a metal gate.

Rate equations in quantum electronics

5.1 Statistics and rate equations

5.1.1 *Introduction*

Rate equations normally are to be interpreted in a statistical sense. In section 1.5 the rate of detection of photons was said to be $dP/dt = p$, but this did not mean that the photons were detected at a uniform rate. There was a Poisson distribution of the detections so that, in one particular interval of duration T, n_1 photons might be detected but in another different interval of the same duration n_2 photons might be detected. The actual numbers of detections fluctuated from interval to interval about the expected value pT. If an ensemble (see section 4.1) of identical systems were considered and detections were made over the same intervals with each system in the ensemble, then an ensemble average would yield the expected value pT. Rate equations provide information about the statistics of ensembles.

Given the close connection between time and energy that is known from quantum theory, it is not too surprising to find that rate equations can also contain statistical information about the distribution of energy amongst the particles. For example, in section 3.2 the rates of change of densities in phase space demonstrated that electrons, regarded as classical particles, were distributed in potential energy Φ according to the Boltzmann distribution

$$n = N_e \exp(-\Phi/kT) \tag{5.1.1}$$

No particular assumptions were made other than the fact that particles were conserved in phase space, so that energy would be conserved automatically.

The Boltzmann distribution is also consistent with the rate equations for recombination given in section 2.3, where

$$dn/dt)_{\text{recombination}} = A(pn - n_i^2) \tag{5.1.2}$$

Here in equilibrium one knows that

$$n = N_c \exp[-(\mathscr{E}_c - \mathscr{E}_f)/kT]$$
$$p = N_v \exp -[(\mathscr{E}_f - \mathscr{E}_v)/kT] \tag{5.1.3}$$

so the electrons and holes obey a similar distribution as (5.1.1) and also satisfy the equilibrium condition for recombination ($dn/dt = 0$) provided that

$$n_i^2 = N_v N_c \exp[-(\mathscr{E}_c - \mathscr{E}_v)/kT] \tag{5.1.4}$$

There is a pleasing consistency that seems to have come about by chance but in fact is part of the consistency that arises from properly constructed rate equations. This initial section demonstrates how the Boltzmann distribution for energy is 'built into' rate equations for interacting particles.

5.1.2 *Classical interactions*

Consider then sets of particles n_q that can exist in any of N_q states which are all at an energy $\mathscr{E}_q$ (Fig. 5.1). An arbitrary range of values of the integer q denotes the different energies. To avoid the difficulties that arise over the occupation of states it is assumed that N_q is so large that nearly all the states are empty. One can then simply argue that there are N_q sites available at energy $\mathscr{E}_q$ to accept a particle. This assumption can later be contrasted with the assumptions for quantum particles. The system is supposed to be a closed system with no

Fig. 5.1. Schematic of sets of energy levels. A change of occupation of one of the n_q particles at energy $\mathscr{E}_q$ to a site at $\mathscr{E}_r$ has to be matched somewhere else by a change of a particle from a site at $\mathscr{E}_j$ to a site at $\mathscr{E}_k$ to conserve energy. Classically $N_q \gg n_q$ (all q) so that effectively there are N_q available sites at each energy.

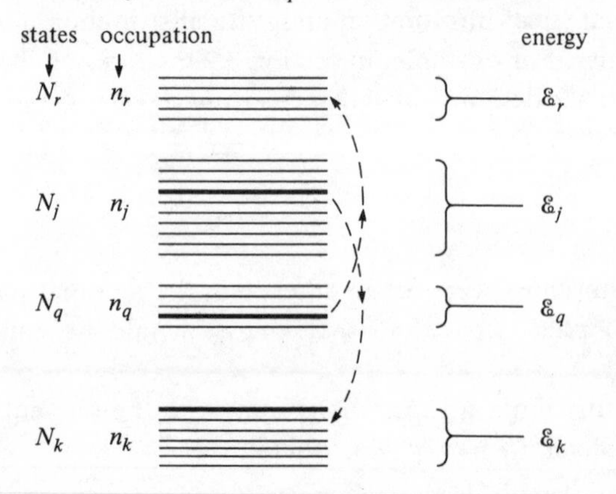

energy lost to or gained from any external source. It is also assumed that the particles interact with only one other particle at any one instant of time.

Suppose then that one of the n_q particles exchanges energy with a similar particle which is one of n_j particles lying in N_j states at energy $\mathscr{E}_j$. The 'q' particle is scattered or changes state to become an 'r' particle in one of the N_r sites at energy $\mathscr{E}_r$. Again N_r is so large that one may regard most of the states as being empty. The 'j' particle with which the 'q' particle interacted, also has to change state in order to conserve energy. So this 'j' particle moves to a 'k' site in one of the N_k states at energy $\mathscr{E}_k$. The initial energy equals the final energy for these interactions requiring that:

$$\mathscr{E}_q + \mathscr{E}_j = \mathscr{E}_r + \mathscr{E}_k \tag{5.1.5}$$

The probability of a 'q' site particle moving to an 'r' site is proportional to the number of particles available to move and to the number of sites available to receive the particle so that

$$\mathrm{d}n/\mathrm{d}t)_{q\,\mathrm{to}\,r} \propto n_q N_r \tag{5.1.6}$$

But the movement from 'q' to 'r' has to be linked, through energy conservation, with movement from 'j' to 'k'. It must again be expected that this last change is proportional to the numbers of 'j' states available to move and the numbers of the 'k' states available to receive the 'j' particles. Consequently it follows:

$$\mathrm{d}n/\mathrm{d}t)_{q\,\mathrm{to}\,r} \propto n_j N_k \tag{5.1.7}$$

Combining the rates of (5.1.6) and (5.1.7) and summing over all possible sets of j and k which satisfy the requirements of conservation of energy yields:

$$\mathrm{d}n/\mathrm{d}t)_{q\,\mathrm{to}\,r} = \sum_{jk} \{A_{qj,rk} n_q n_j N_r N_k\} \tag{5.1.8}$$

where $A_{qj,rk}$ is not known here but is fixed by the detailed physical laws of the particular interaction that is being considered.

(5.1.8) gives just a portion of the interactions which remove particles from the energy state 'q'. A summation over all values of r gives the total rate of removal from the state 'q'.

A fundamental concept in these rate equations is that the term $A_{qj,rk}$ represents the probability of a change per *single* particle at 'q' to a single vacant site at 'r', combined with a movement of a single particle from 'j' to a single site at 'k'. This probability is said to be the same as the basic probability per particle per vacant site for going from 'r' to 'q' accompanied by a change from 'k' to 'j'. Exactly opposite motions are believed to have the same *a priori* probability, and it is the relative numbers of available particles that finally fix the distribution. It is asserted then that

$$A_{qj,rk} = A_{rk,qj} \tag{5.1.9}$$

The justification for (5.1.9) is made by considering what happens by reversing time for a cloud of interacting particles where all energy is accounted for in the interactions and no energy is transported away from the system. Suppose a film was made a long time after the system had settled down to 'equilibrium'. Any artificial initial order would have been removed. An audience who then saw a section of the film running backwards would not be able to tell that the process had been reversed. simple elastic collisions between billiard balls show this feature, and it is supposed that this 'time reversal' property carries over into all types of interaction. (For some types of particles in high energy physics time reversal[48,49] is not thought to hold. It appears that neutral kaon particles decay into two pions and violate time reversal. However, this need not directly concern most electronic physicists and engineers.)

Scattering into 'q' is then formed in a similar way to scattering out (see (5.1.8)) to give the total scattering into and out of 'q' as:

$$\begin{aligned} \mathrm{d}n_q/\mathrm{d}t &= \sum_{r,j,k} \{A_{qj,rk}[N_qN_jn_rn_k - n_qn_jN_rN_k]\} \\ &= \sum_{r,j,k} \{A_{qj,rk}n_qn_jn_rn_k[(N_q/n_r)(N_j/n_j) - (N_r/n_r)(N_k/n_k)]\} \end{aligned} \tag{5.1.10}$$

The triple summation is made in (5.1.10) with the restriction that energy is conserved (see (5.1.5)).

In equilibrium $\mathrm{d}n_q/\mathrm{d}t = 0$ over an ensemble of systems. The *detailed balance* that is required to bring this equilibrium about for all systems regardless of the precise mechanisms of interaction is accomplished by having $(N_q/n_q)(N_j/n_j) = (N_r/n_r)(N_k/n_k)$ or equally

$$\ln(N_q/n_q)\ln(N_j/n_j) = \ln(N_r/n_r) + \ln(N_k/n_k) = f \tag{5.1.11}$$

Comparison of (5.1.5) and (5.1.11) shows that f must be a function $f(\mathscr{E})$, with $f(\mathscr{E})$ and $\mathscr{E}$ both conserved in a collision. The principle of detailed balance used in (5.1.11) then assures equilibrium simultaneously with energy conservation.

The functional dependence of n_q upon $\mathscr{E}_q$ can be tested by comparison of (5.1.5) and (5.1.11). Remembering that these hold for all q, j, r, k, it follows that

$$\ln(N_q/n_q) = b(\mathscr{E}_q - \mathscr{E}_\mathrm{R}) \tag{5.1.12}$$

with $\mathscr{E}_\mathrm{R}$ and b independent of the state q; then one expects the occupation of an energy state, averaged over members of an ensemble of identical systems to be given by

$$n_q = N_q \exp[-b(\mathscr{E}_q - \mathscr{E}_R)] \tag{5.1.13}$$

The sign of b has been taken so that the probability decreases as energy increases, consistent with elementary physical concepts.

Rate equations do not in general give the values of the constants which are used, and these constants have to be determined from experiment or from comparison with other theories. Comparison with an ideal perfect gas gives

$$b = 1/kT \tag{5.1.14}$$

where k is Boltzmann's constant and T is the temperature. The constant a is determined from the total numbers N:

$$N = \sum_q n_q = \sum_q N_q \exp[-b(\mathscr{E}_q - \mathscr{E}_R)] \tag{5.1.15}$$

$$\mathscr{E}_R = kT \ln\left\{N \middle/ \left[\sum_q N_q \exp(-\mathscr{E}_q/kT)\right]\right\} \tag{5.1.16}$$

The energy $\mathscr{E}_R$ provides a reference energy determined by the states and the numbers of particles.

(5.1.16) determines a definite value for $\mathscr{E}_R$ when N is known. Typically, in electronics or physics, the numbers of electrons or holes are fixed. In these cases this reference energy is referred to as the 'Fermi' energy: $\mathscr{E}_R = \mathscr{E}_f$. However, in the absence of any restriction on numbers, the restriction on $\mathscr{E}_R$ must be that the expected numbers add up to the whole so $\exp(\mathscr{E}_R/kT) = 1$ or $\mathscr{E}_R = 0$. (5.1.16) is then consistent for all N regardless of the value of temperature T.

The result of (5.1.14)-(5.1.16) gives the classical Maxwell-Boltzmann equilibrium probability distribution of particles[48,50,51] (Fig. 5.2) with

Fig. 5.2. The Maxwell-Boltzmann energy distribution. ($kT \sim 25$ meV at room temperature.)

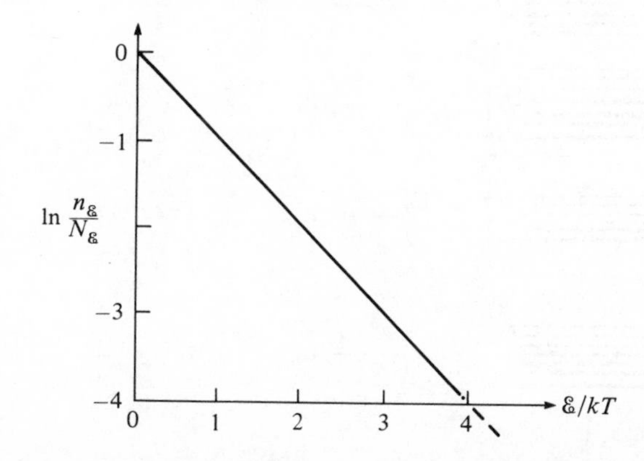

energy:

$$\text{MB}(\mathscr{E}) = n_{\mathscr{E}}/N_{\mathscr{E}} = \exp[-b(\mathscr{E} - \mathscr{E}_{\text{R}})] \tag{5.1.17}$$

This is the distribution found earlier in studying the Boltzmann equation and semiconductor transport. It may be thought from the derivation that it applies only if particles interact in pairs, but it is possible to show that the same result has to hold for interactions with three or more particles, and an alternative method of calculation by considering the most probable arrangements of particles also gives the same result.[48] The remaining task of finding the density of states $N(\mathscr{E})$ is left to appendix A and to further reading.[5,8,9] The next question that arises is 'how does quantum theory change these statistical interactions?'

5.2 Fermion–fermion interactions

Electrons and holes are examples of 'fermion' quantum particles which obey the 'Pauli exclusion principle', namely that not more than one particle may occupy precisely the same quantum state.[52] The previous discussion avoided this issue by arguing that the number of states was so much larger than the number of particles trying to fill them that it was not necessary to consider how the particles filled the states. This section extends the discussion to include the Pauli quantum restriction.

The scene is the same as previously in that there is a closed system with no energy being lost to or gained from any external source. Within

Fig. 5.3. Schematic for interchange of fermions. System as for Fig. 5.1, but note that vacant sites are $(N_r - n_r) > 0$, and if there is no vacant site then $(N_r - n_r) = 0$ and particles cannot move to that site.

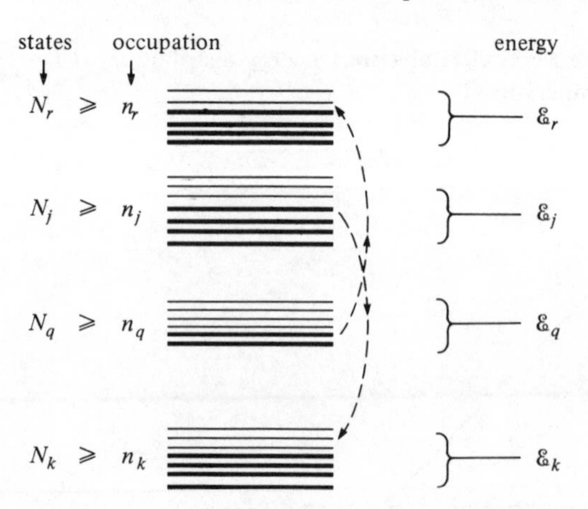

this system there are n_q particles in N_q quantum states (see appendix A) all at energy $\mathscr{E}_q$ (Fig. 5.3). There are numerous such sets of particles which interact with one another. In an interaction, again only two particles at one instant in time are assumed to be involved so that one 'q' state particle interacts with one 'j' state particle. The 'q' state moves to an 'r' state while, to conserve energy, the 'j' state moves to a 'k' state ((5.1.5) relating initial and final energies still holds). The probability of this movement happening for just one particle at 'q' and at 'j' with one vacancy at 'r' and at 'k' is given by $A_{qj,rk}$ as before. However, this single particle probability $A_{qj,rk}$ applies to each of the n_q particles so that the overall probability is increased by the number of particles n_q. Now each 'q' particle has $(N_r - n_r)$ possible vacant sites at 'r' to which they may transfer, so that the probability must be further increased by the available number of states at 'r', namely $(N_r - n_r)$ (>0). This last result uses the Pauli principle so that if $N_r = n_r$ there is no possibility of particles transferring to the 'r' state. The related movement from 'j' to 'k' in order to conserve energy is also dependent upon the numbers n_j available to move and on the number of vacant sites $(N_k - n_k)$ (>0) available to receive the particle from 'j'. The movement out of 'q' is then determined by

$$\left.\mathrm{d}n_q/\mathrm{d}t\right)_{\text{from } q \text{ to } r} = \sum_{jk} \{A_{qj,rk}[n_q n_j (N_r - n_r)(N_k - n_k)]\} \tag{5.2.1}$$

As in the previous work, time reversal assures one that $A_{qj,rk} = A_{rk,qj}$ so that the total scattering both into and out of 'q' is given by

$$\mathrm{d}n_q/\mathrm{d}t = \sum_{r,j,k} \{A_{qj,rk}[n_r n_k (N_q - n_q)(N_j - n_j) - n_q n_j (N_r - n_r)(N_k - n_k)]\}$$

On rearranging, by writing

$$\begin{aligned} R_m &= (N_m/n_m) - 1 \\ \mathrm{d}n_q/\mathrm{d}t &= \sum_{r,j,k} \{A_{qj,rk} n_q n_j n_r n_k [R_q R_j - R_r R_k]\} \end{aligned} \tag{5.2.2}$$

The initial energy and final energy are unchanged for each interaction so again

$$\mathscr{E}_q + \mathscr{E}_j = \mathscr{E}_r + \mathscr{E}_k = \mathscr{E} \tag{5.2.3}$$

In equilibrium with $\mathrm{d}n_q/\mathrm{d}t = 0$, using the principle of detailed balance

$$\ln R_q + \ln R_j = \ln R_r + \ln R_k = f \tag{5.2.4}$$

From functional similarity of (5.2.3) and (5.2.4) it is seen that $f = f(\mathscr{E})$ so that both $f(\mathscr{E})$ and $\mathscr{E}$ are conserved at a collision or interaction (see (5.1.5) and (5.1.11)).

The functional dependence of R_q on $\mathscr{E}_q$ follows from (5.2.3) and (5.2.4) (see (5.1.12)):

$$\ln R_q = b^+(\mathscr{E}_q - \mathscr{E}_f) \tag{5.2.5}$$

where b^+ and $\mathscr{E}_f$ have to be determined by a separate experiment or calculation but are independent of the value of q. In the limit where there are very few particles ($n_q \ll N_q$), there can be no difference between the classical and this quantum mechanical version, so it follows that $b^+ = b = 1/kT$, independent of energy, as before. The equivalence of b and b^+ can be checked also by considering the interaction between classical particles and fermions (see problem 5.5).

The final result for the distribution with energy is then given as the Fermi–Dirac distribution (Fig. 5.4):

$$F(\mathscr{E}) = n_q/N_q = 1/\{1+\exp[(\mathscr{E}_q - \mathscr{E}_f)/kT]\} \tag{5.2.6}$$

The 'Fermi' energy $\mathscr{E}_f$ is found from the total numbers

$$N = \sum_q \{n_q\} \tag{5.2.7}$$

It is seen that it does not matter how weak is the interaction between the fermions, they still settle down after enough time to this Fermi–Dirac distribution. Any closed system of fermion particles, which are in equilibrium, then has to have the same Fermi energy level $\mathscr{E}_f$ throughout the system.

5.3 Photon–electron interactions

In the previous account of how a group of electrons reached equilibrium there was no discussion of any mechanism that could help electrons in one region interact with electrons some distance away. The concept of the electromagnetic field associated with the electrons means that, to exchange energy, the electron in the 'q' state gives out a quantum

Fig. 5.4. The Fermi–Dirac distribution $F(\mathscr{E})$. $\mathscr{E}_f$ is the Fermi energy, where $F(\mathscr{E}_f) = \frac{1}{2}$ determined by numbers of electrons and available sites.

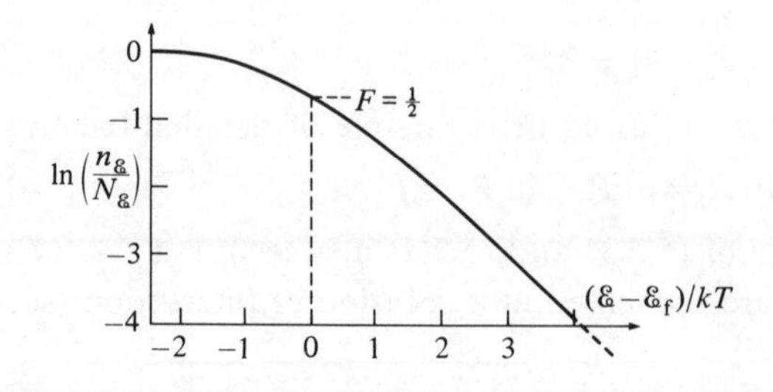

of energy $\mathscr{E}_o$ to the field as the electron falls to a lower energy 'r' state.

$$\mathscr{E}_o = (\mathscr{E}_q - \mathscr{E}_r) \quad (5.3.1)$$

A 'j' state electron then may or may not reabsorb this quantum of energy $\mathscr{E}_o$, but if it does absorb the energy then the 'j' electron moves up to one of the empty or available 'k' states where $\mathscr{E}_k - \mathscr{E}_j = \mathscr{E}_o$. This section addresses the problem of energy emitted and absorbed by the electromagnetic field.

There is an implicit assumption that the quantum of energy given by the interaction of electrons with vacant sites is accomplished by one 'particle' - a photon - being associated with any one interaction. An electron which falls in energy then gives out a photon to the electromagnetic field. An electron which gains energy absorbs a photon. This description of the interaction of electrons and the electromagnetic field is valid for all frequencies, but it is experimentally observed most clearly for those frequencies where the quantal energy (hf) is in excess of the thermal energy (kT), as, for example, with infra-red, visible optics, x-rays and gamma-rays.

The discussion follows closely, but in a more serious way, the discussion in chapter 1 over buying and selling houses. The detailed system that is considered here is shown in Fig. 5.5, where a uniform unit volume of material is considered with N_c electron quantum states at an energy $\mathscr{E}_c$ containing n_c ($<N_c$) electrons. At a lower energy $\mathscr{E}_v$ there are n_v electrons in the N_v states giving $(N_v - n_v)$ vacant sites. The electromagnetic field is assumed to have P quanta each of energy $\mathscr{E}_o$ which can exchange energy with the electrons. The unit of volume is assumed to be sufficiently uniform that no spatial changes of the interaction need be considered.

Fig. 5.5. Schematic diagram for electron-photon interactions in a closed system. Only two energy levels, $\mathscr{E}_c$ and $\mathscr{E}_v$, containing N_c, N_v sites with n_c, n_v electrons - approximates to system in semiconductor with conduction band and valence band electrons and density P of photons. (a) Spontaneous transitions. (b) Stimulated absorption. (c) Stimulated emission.

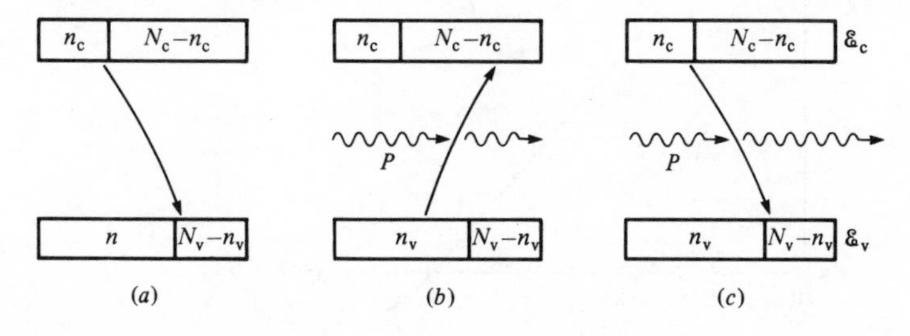

The problem of conservation of momentum was discussed briefly in Figs. 2.5 and 2.6, and the same considerations apply here so that the model of Fig. 5.5 resembles a direct band gap semiconductor with the conduction band carrying $n = n_c$ electrons and with N_c states all effectively lumped together into one energy level. Similarly all the valence band states N_v are lumped together with $p = N_v - n_v$ giving the holes. A more detailed discussion of this aspect follows in chapter 6.

For the present assume that this uniform system is entirely self contained so that no energy is lost from the system and therefore all the photons remain and interact with the electrons. We consider the rates of change of the photon density P.

First, if any electron falls spontaneously in energy to the lower site then a photon will be given out. For one isolated electron and one isolated vacant site the probability of such an emission in time δt is $A\delta t$. However, there are n_c electrons available to give out energy, and each one of these n_c electrons can fall into one of $(N_v - n_v)$ vacant sites available to receive the electron so that the probability A of emission per unit time is increased by $n_c(N_v - n_v)$, giving:

$$\mathrm{d}P/\mathrm{d}t)_{\text{spontaneous}} = An_c(N_v - n_v) \tag{5.3.2}$$

The spontaneous emission has already been met in the discussion of recombination. Electrons are able to fall spontaneously from a higher to a lower energy provided at the lower energy there is a hole with which it may recombine leading to a recombination rate that is proportional to np.

Fig. 5.6. The Bose–Einstein distribution.

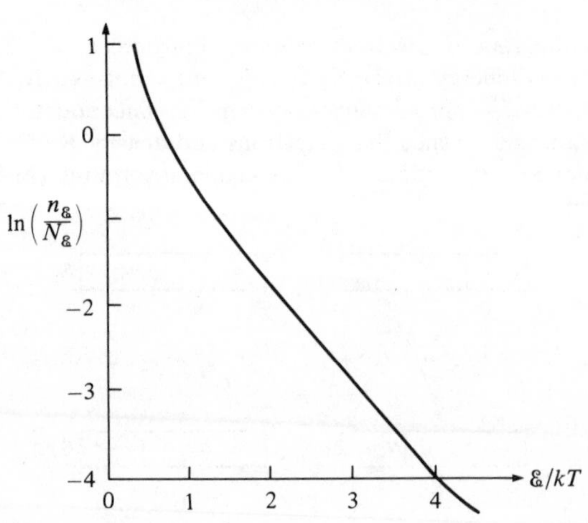

To go from energy $\mathscr{E}_v$ up to $\mathscr{E}_c$, the electron requires energy $\mathscr{E}_o$; given here by a photon. The number of electrons available to receive energy at $\mathscr{E}_v$ is n_v, while the number of vacant sites available to receive the electron at the higher energy are $(N_c - n_c)$ and there are P photons to give up energy. If the probability per photon per available electron per vacant site is B per unit time, then it follows that the loss of photons stimulated by absorption is given by

$$\mathrm{d}P/\mathrm{d}t)_{\text{stimulated absorption}} = -BPn_v(N_c - n_c) \qquad (5.3.3)$$

If, at this stage, the stimulated absorption was balanced against the spontaneous emission then one would find that in equilibrium

$$P = (A/B)\exp -(\mathscr{E}_o/kT)$$

giving a classical Maxwell-Boltzmann distribution!

Now, it is again argued using the general principle of time reversal that for the process of stimulated absorption there is a converse process of stimulated emission, with the same coefficient of probability B.† There are still P photons available at the energy $\mathscr{E}_o$ to stimulate the emission of an electron. But now there are n_c electrons each with the same probability of being stimulated to give out additional energy into the photons and each able to fall into one of $(N_v - n_v)$ available vacant sites. Consequently one arrives at the probable rate

$$\mathrm{d}P/\mathrm{d}t)_{\text{stimulated emission}} = BPn_c(N_v - n_v) \qquad (5.3.4)$$

Collecting all the terms together for the closed system:

$$\mathrm{d}P/\mathrm{d}t = An_c(N_v - n_v) + BP[n_c(N_v - n_v) - n_v(N_c - n_c)] \qquad (5.3.5)$$

This leads to the equilibrium value $(\mathrm{d}/\mathrm{d}t = 0)$

$$P = (A/B)R_v/(R_c - R_v)$$

where using (5.2.6)

$$R_{c/v} = (N_{c/v}/n_{c/v}) - 1 = \exp[(\mathscr{E}_{c/v} - \mathscr{E}_f)/kT]$$

The result is independent of the Fermi level of the electrons in the material with

$$P = (A/B)B(\mathscr{E}); \quad B(\mathscr{E}) = 1/[\exp(\mathscr{E}/kT) - 1] \qquad (5.3.6)$$

$B(\mathscr{E})$ is known as the Bose-Einstein distribution (Fig. 5.6).[50,51]

The rate equations cannot by themselves give the values of the 'Einstein coefficients' A and B nor can they give the ratio A/B of the spontaneous

† Footnote: time reversal for spontaneous emission does not apply because this would require an electron to rise in energy without gaining energy from the P available photons in the field. Time reversal arguments are closely linked with conservation of energy which gives the symmetry in time according to quantum theory.

probability to the stimulated probability. Detailed quantum theory shows that this ratio depends on the number of modes of the electromagnetic field which can support the photons of energy $\mathscr{E}_o$. The calculation above is done for photons per unit volume so the total number of photons are $P_t = P\Phi$, where Φ is the interaction volume. For photons in a single mode it is found that $A\Phi = B$. Appendix A gives a standard calculation for the density of states A/B for 'black body' radiation.

The Bose–Einstein distribution is relevant to the 'boson' family of particles of which the photon is but one example. All classical harmonic oscillators, with a vibrational or oscillation frequency ω and energy proportional to the square of the amplitude of oscillation, quantise into boson particles such that energy is exchanged in units of $\hbar\omega$. For example, phonons are the boson particles associated with the lattice vibrations of a crystal.

A key difference between bosons and fermions is that there can be any number of bosons in a single quantum state of energy $\mathscr{E}_o$. The electromagnetic field can continue to be raised to higher values of energy through the emission of photons of the same energy. The Pauli principle does not apply to bosons. In high energy physics there are a plethora of boson particles, but these are again of little immediate importance for the practical electronic physicist or engineer. Liquid helium particles can condense into boson states, and this condensation in which large numbers of particles can be in identical quantum states gives the features of superfluidity. Coupled pairs of fermions can act almost as bosons. This has important effects in superconductivity, which is a result of large numbers of pairs of electrons all cooperating in a single quantum state giving large currents at zero voltage drop.

An alternative interpretation of the result of (5.3.5) can be seen by multiplying the whole equation by the interaction volume Φ and writing P now as the *total* number of photons. For the single mode, where $A\Phi = B$

$$dP/dt = B[(P+1)n_c(N_v - n_v) - Pn_v(N_c - n_c)] \tag{5.3.7}$$

The stimulated emission in a single mode is proportional to $P+1$, while the absorption is proportional to P. The extra 1 in the stimulation process is sometimes said to be caused by photons in the 'quantum ground state', which are not available to give up energy but can randomly stimulate electrons into giving up energy – spontaneous emission. The ground state cannot stimulate absorption, but causes fluctuations in the field, which leads to noise, equivalent to an energy of $\frac{1}{2}\hbar\omega$ per mode.[53,54] However, unlike the thermally induced noise of fluctuations, these quantum fluctuations are fundamental to the giving of noise limits for optical communication systems.

In practical devices one expects photons to be able to enter an interaction region and generated photons to escape from that region. A characteristic time (the photon lifetime) τ_p can be assigned to the time scale of the loss and gain so that

$$\mathrm{d}P/\mathrm{d}t)_{\text{external}} = -(P - P_e)/\tau_p \tag{5.3.8}$$

When the material is in equilibrium with the external electromagnetic field, the photons leaving are on average balanced by the photons arriving; the magnitude of P_e must be this equilibrium value.

The result of (5.3.5) can be combined with the result of (5.3.8). It is also helpful, considering semiconductors, to put (5.3.5) into a tidy form by writing in terms of density of electrons n in the equivalent conduction band and density of holes p in the equivalent valence band for a unit volume of material:

$$n_c = n, \quad p = N_v - n_v$$

Hence

$$\mathrm{d}P/\mathrm{d}t = BPN_cN_v[(p/N_v) + (n/N_c) - 1] + Bnp - (P - P_e)/\tau_p \tag{5.3.9}$$

Here the first term gives the net stimulated emission (if the values of n and p give a net positive output) or absorption (negative net emission). The second term Bnp gives the spontaneous emission, while the final term gives the loss of photons to the external regions of the material, and these are the potentially useful photons that can give out light from a p–n junction, for example.

The discussion may be advanced by using the Fermi–Dirac distribution $F(\mathscr{E})$ for electrons and writing

$$n = N_c/\{1 + \exp[(\mathscr{E}_c - \mathscr{E}_{fn})/kT]\}$$
$$p = N_v/\{1 + \exp[-(\mathscr{E}_v - \mathscr{E}_{fp})/kT)\}$$

where $\mathscr{E}_{fn}$ is the effective Fermi level or 'imref' used to describe the non-equilibrium value of electron density, and similarly $\mathscr{E}_{fp}$ is the effective Fermi level for the non-equilibrium density of holes. The condition is required that the stimulated emission term (the coefficient of BP in (5.3.9)) be positive:

$$1 > \exp\{(\mathscr{E}_c - \mathscr{E}_v) - (\mathscr{E}_{fn} + \mathscr{E}_{fp})]/kT\}$$

or

$$(\mathscr{E}_{fn} - \mathscr{E}_{fp}) > (\mathscr{E}_c - \mathscr{E}_v) \tag{5.3.10}$$

Net stimulated emission demands then a highly non-equilibrium condition in a semiconductor whereby the effective Fermi levels for the electrons and the holes are separated by the application of a bias voltage in excess of the band gap voltage. Practical experience will tell one that normally for p–n junction diodes the forward voltage, for a reasonable practical current density lies typically around $\frac{3}{4}$ of the band gap voltage (e.g. 0.7 V for a Si diode) and the current density could increase dramatically to well over 10^3 A/cm^2 as the forward bias was increased to the band gap value! In spite of the apparent simplicity of stimulated emission in a p–n junction this hurdle of a high current density has been a major difficulty that has had to be overcome by careful design.

Rate equations will be discussed for diode semiconductor lasers in a more detailed way at a later stage.

5.4 More about bosons

Having introduced bosons, it is helpful to consider the interactions between sets of bosons alone and to examine the rate equation treatment comparable to that given for fermions. With fermions the important distinction in the rate equations was that not more than one particle could be packed into a quantum state. In quantum theory, bosons are 'particles' which have no restrictions on how many can be packed into a single boson quantum state.

Consider then as before n_q particles in N_q quantum states at energy $\mathscr{E}_q$ within a closed system (Fig. 5.1). In an interaction, again only two particles at a time are assumed to be involved so that one 'q' state particle interacts with a 'j' state particle. The 'q' state moves to an 'r' state while the 'j' state moves to a 'k' state to conserve energy. The probability of this event happening for just one particle at 'q' and at 'j' with one vacancy at 'r' and at 'k' is given by $B_{qj,rk}$.

Now in considering the n_q bosons, each of these can interact with any one of the n_j bosons, but there has to be a suitable set of sites into which these particles can scatter. Figure 5.7 shows an example of the problem with $n_q = 11$ bosons and $N_q = 5$ state cells all at the same energy $\mathscr{E}_q$. It can be seen from the specific example that, with no restrictions on the number in each cell, the next boson to arrive may be placed at any position in the ring. There are $16 = (N_q + n_q)$ such positions. The result holds for all values of q, r, etc. Therefore, when it comes to considering the probability of transferring a 'q' boson into an 'r' state, each of the n_q bosons available to transfer has $(N_r + n_r)$ available sites into which it can go at the energy $\mathscr{E}_r$. The probability is therefore increased by the factor $(N_r + n_r)$ as compared to N_r for classical particles and as compared to $(N_r - n_r)$ for fermions. It may also be helpful to compare this discussion with that of the preceding section for fermion–boson interactions. It is then possible to consider splitting the multiplier $(N_r + n_r)$ into 'spontaneous' exchanges proportional to N_r and 'stimulated' exchanges proportional to n_r.

To conserve energy, this transfer has to be accompanied by a transfer of one of the n_j bosons into one of the (N_k+n_k) available 'k' sites. The overall probability of scattering out is then determined by

$$\mathrm{d}n_q/\mathrm{d}t)_{\text{from } q \text{ to } r} = \sum_{jk} \{B_{qj,rk} n_q n_j (N_r+n_r)(N_k+n_k)\} \tag{5.4.1}$$

Once again time reversal arguments suggest that $B_{qj,rk} = B_{rk,qj}$ so that the total scattering both into and out of 'q' is given by

$$\mathrm{d}n_q/\mathrm{d}t = \sum_{r,j,k} \{B_{qj,rk}[n_r n_k (N_q+n_q)(N_j+n_j) - n_q n_j (N_r+n_r)(N_k+n_k)]\} \tag{5.4.2}$$

On rearranging, by writing

$$S_m = (N_m/n_m)+1$$

$$\mathrm{d}n_q/\mathrm{d}t = \sum_{r,j,k} \{B_{qj,rk}[n_q n_j n_r n_k (S_q S_j - S_r S_k)]\} \tag{5.4.3}$$

The initial energy and final energy are unchanged for each interaction so again

$$\mathscr{E}_q + \mathscr{E}_j = \mathscr{E}_r + \mathscr{E}_k = \mathscr{E} \tag{5.4.4}$$

In equilibrium with $\mathrm{d}n_q/\mathrm{d}t = 0$, using detailed balancing as before

$$\ln S_q + \ln S_j = \ln S_r + \ln S_k = f(\mathscr{E}) \tag{5.4.5}$$

With $f(\mathscr{E})$ and $\mathscr{E}$ both conserved at a collision or interaction.

Fig. 5.7. Calculation for finding rate at which next boson can enter energy level $\mathscr{E}$. There are no limitations on how many bosons are in each state. Example shows 11 particles and 5 states, which are all at the same energy. This gives 16 different slots into which the next boson can be placed.

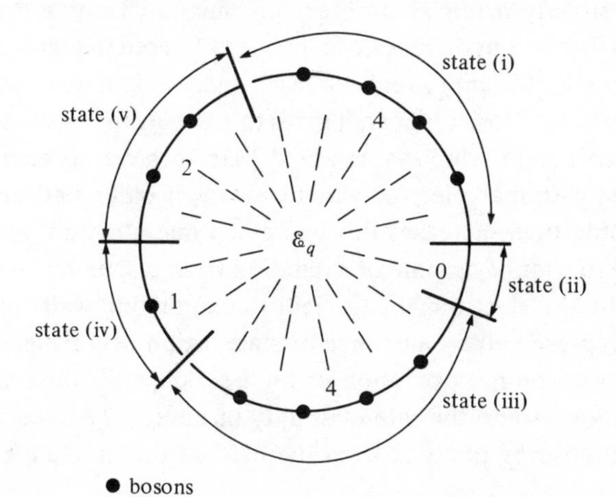

The functional dependence of S_q on $\mathscr{E}_q$ is found as before by comparison of (5.4.4) and (5.4.5), noting that these equations hold for all q, r, j and k. Thus

$$\ln S_q = b^{++}(\mathscr{E}_q + \mathscr{E}_R) \tag{5.4.6}$$

where $\mathscr{E}_R$ is some reference energy and b^{++} again must be consistent given in the limit when $N_q \gg n_q$, so $b^{++} = b = 1/kT$, the classical value independent of energy.

The final result for the distribution with energy is then given as

$$n_q = N_q/\{\exp[(\mathscr{E}_q + \mathscr{E}_R)/kT] - 1\} \tag{5.4.7}$$

The reference energy $\mathscr{E}_R$ is found from the total numbers

$$N = \sum_q \{n_q\} \tag{5.4.8}$$

Liquid helium can behave at low enough temperatures as a set of bosons, and the number of atoms is clearly limited so that (5.4.8) is relevant to this problem. However, from the discussion of interaction of bosons and fermions, it can be seen that, for photons where the numbers are not limited, the reference energy $\mathscr{E}_R$ has to be zero to give consistent results at equilibrium and for interaction with electrons. So for photons with an energy $h\nu = \mathscr{E}_q$ the result of (5.3.6) holds, with

$$n_q = N_q B(\mathscr{E}), \quad \text{where } B(\mathscr{E}) = 1/[\exp(\mathscr{E}_q/kT) - 1] \tag{5.4.9}$$

The Bose–Einstein distribution is recovered again (Fig. 5.6).

5.5 Time fluctuations of photons in the electromagnetic field

In the introduction to this chapter it was indicated that temporal statistical information about the emission of photons could be obtained from rate equations. This task is briefly dealt with here. The discussion is limited to a slightly artificial problem of emission from a thermal source giving an incoherent or chaotic field[54] but filtered to give a single mode or wavelength. The more realistic laser analysis is left for later.

Each time an electron loses energy, it gives this energy to the surrounding electromagnetic field which is assumed here to have an energy U and be contained within an electromagnetic cavity or other well-defined region. The photon then increases this energy by one quantum and the field can be said to change its state of excitation from a state U_n to U_{n+1}. Loss of energy by the electromagnetic field to conducting walls or to a load is equally represented by a change of state, again occurring at one quantum at a time. The net excitation of the field depends then on the rate of loss of photons and the rate of supply of photons. A receiver of an optical signal absorbs photons from the field so that fluctuations in the field appear as noise on the received signal. If the optical source is an 'ideal' laser then the photons are emitted into the field with a Poisson distribution and the fluctuations can be calculated (see problem 1.4). However, the statistical fluctuations of a more realistic laser are complex, and it is easier to make a start with a model which is appropriate for the generation of incoherent light at a single optical mode.

(5.3.7), the equation for the total emission into a single optical mode, is rewritten in the form

$$\mathrm{d}P/\mathrm{d}t = B(P+1)[np] - BP[(N_v - p)(N_c - n)] - (P/\tau) \qquad (5.5.1)$$

The first term on the r.h.s. is the photon emission and, as indicated in the preceding section, is proportional to $(P+1)$ for P photons present in this single mode. The next term is the absorption and is proportional to the number of photons P. The final term gives the rate of loss of photons to the external field. For this present calculation the equilibrium value of P_e inserted previously is regarded as negligibly low and is ignored here. The time τ is referred to as the photon lifetime, and for many semiconductor devices will be measured in picoseconds and is taken to be independent of the optical intensity.

For a laser, the optical field stimulates a coherent action over a relatively large volume for the emission or absorption of photons by the electrons, and we ignore spatial variations of the emission and absorption in this discussion. The electron density then drops if the light output increases and, equally, rises if the light output drops. Thus there is a feedback in the lasing mechanisms which helps to stabilise the fluctuations in the photon output. However, for an incoherent output there are some electrons absorbing and some emitting so that one needs only to consider the average value of n and p, which change on a slow time scale compared with the emission and absorption processes. Indeed for simplicity here n and p are taken as constant for the calculation, and the rate equations determine the fluctuations in the excitation state of the field.

To shorten writing put the optical gain rate as

$$g = Bpn \qquad (5.5.2)$$

where, for a light emitting diode or a semiconductor laser, $1/g$ is measured in tens or hundreds of femtoseconds. The absorption rate can also be abbreviated to

$$B(N_c - n)(N_v - p) = \alpha g \qquad (5.5.3)$$

Define E_P as the probability that the optical field is in the Pth excited state with P photons. Given the field is in the Pth state then there is a probability of $g(P+1)$ for stimulated emission and of αgP for absorption.

The probable rate of loss from the region is given by P/τ. Thus adding all these probabilities, multiplied by the initial probability E_P of being in the 'P' state, gives the overall rate of reduction of the probability E_P as:

$$\mathrm{d}E_P/\mathrm{d}t)_{\text{reduction}} = -E_P[g(P+1)+\alpha gP+(P/\tau)] \tag{5.5.4}$$

Similarly if the state is already in the '$P-1$' state then the probability of an emission is $(P-1+1)g = Pg$, bringing the state back into a P state. This increases the probability of E_P at a rate $E_{P-1}Pg$. If the state is already in a '$P+1$' state then the probability of an absorption is $\alpha g(P+1)$ and the probability of loss from the region is $(P+1)/\tau$, both processes bringing the state back into a 'P' state. Thus there is a further rate of increase in the probability E_P given from $E_{P+1}(P+1)[\alpha g+(1/\tau)]$. Combining these rates of gain with the rates of reduction gives the overall rate of change for the probability E_P as

$$\begin{aligned}\mathrm{d}E_P/\mathrm{d}t &= -E_P[g(P+1)+\alpha gP+(P/\tau)]\\ &\quad + E_{P+1}(P+1)[\alpha g+(1/\tau)] + E_{P-1}Pg\\ &= (P+1)\{E_{P+1}[\alpha g+(1/\tau)] - E_P g\}\\ &\quad - P\{E_P[\alpha g+(1/\tau)] - E_{P-1}g\}\end{aligned} \tag{5.5.5}$$

In equilibrium with $\mathrm{d}E_P/\mathrm{d}t = 0$ for all P, it follows that

$$E_{P+1}/E_P = g\tau/(1+g\alpha\tau) = F \tag{5.5.6}$$

or

$$E_P = E_0 F^P \tag{5.5.7}$$

However, if E_P is to be correctly interpreted as a probability then it must follow that

$$\sum_{P=0}^{\infty}\{E_P\} = 1; \quad \sum_{P=0}^{\infty}\{PE_P\} = Q \tag{5.5.8}$$

where Q is the average value of P.

$$E_P = F^P(1-F) \quad \text{with } F/(1-F) = Q$$

so that

$$E_P = Q^P/(1+Q)^{1+P} \tag{5.5.9}$$

By summing a geometric series it may be checked that the average value $\langle P\rangle = Q$ and, further, that

$$\langle P^2\rangle = \sum_{P=0}^{\infty}\{P^2E_P\} = 2\langle P\rangle^2 + \langle P\rangle \tag{5.5.10}$$

If the emission of the incoherent photons had been a Poisson distribution then

$$\langle P^2 \rangle = \langle P \rangle^2 + \langle P \rangle \tag{5.5.11}$$

It will be seen that incoherent emission is statistically more noisy than the coherent laser emission.

Contrasting the result for single mode chaotic light of (5.5.9) with laser light, which is known to have a Poisson distribution:

$$E_P = (Q^P/P!) \exp(-Q) \tag{5.5.12}$$

Fig. 5.8 contrasts the two different arrival rates and problem 5.6 considers one important implication.

The reader may have previously understood, not incorrectly, that even thermal light is usually emitted with a Poisson distribution. Indeed, in any system where on average the output P is determined by the rate $dP/dt = p$ with p a constant, then the output will be a Poisson distribution. This result which refers to the output of several spontaneous modes of emission emitting together is not at variance with the result of (5.5.9), which refers to photon fluctuations in a *single* photon mode. Averaged over many modes which are the characteristics of normal thermal light one arrives back at the Poisson distribution (see problem 5.3). The point to make here is that if one could filter thermal light into as narrow a band of frequencies as a laser then would discover that there was a significant statistical difference between the laser light and the thermal light. The line-width would not be sufficient measure of the difference between the two types of source; laser light and chaotic light are then fundamentally different in quality and that quality can be measured by photon statistics. If lasers did not exist and one could only filter chaotic

Fig. 5.8. Photon probabilities for chaotic distribution and ideal. Poisson distribution (laser light) in a single mode. (Comparison for mean of 15 photons.)

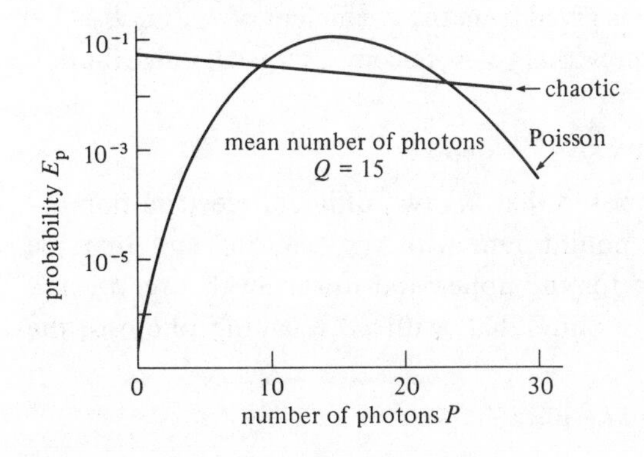

light, then the additional fluctuation noise might degrade high quality digital optical communication systems where ideally one requires an error rate of less than 1 in 10^9.

It will also be seen later, from chapter 6, that $g(1-\alpha)=1/\tau$ is the threshold condition for lasing in the semiconductor material. This condition therefore gives $F=1$ ($Q=\infty$) at which the above theory clearly breaks down because a steady output is not possible. The statistics of laser light must be left for a further chapter.

PROBLEMS 5

5.1 Find the wavelength λ and frequency ν at which $kT=h\nu$ with kT equal to room temperature (300 K), liquid nitrogen (77 K), and liquid helium (4 K).

5.2 Set up a rate equation for photons at energy $\hbar\omega_1$, $\hbar\omega_2$ to produce photons at energy $\hbar\omega_{12}=\hbar\omega_1-\hbar\omega_2$ in a single mode system, remembering that emission is proportional to $P+1$ and absorption proportional to P for each component taking part. Check that each equation at equilibrium satisfies the requirement that the photons satisfy Bose–Einstein statistics. Hence show that if power is fed in at ω_1 it is given out at ω_{12} and ω_2 according to

$$\text{power}_{12}/\omega_{12}=\text{power}_2/\omega_2=\text{input power}_1/\omega_1$$

These relations are known as the Manley–Rowe relations.

5.3 Consider a set of chaotic distributions where

$$E_P=E_0F^P$$

so that the probability of the P photon state (E_P) is the coefficient of F^P in $E_0/(1-F)$. Let the gain be reduced for N such independent systems all feeding into the same output. The new value of 'F' is then F/N in each system. Show that the probability of the P photon state is given from the coefficient of F^P in $[E_0/(1-F/N)]^N$. Hence show that as N becomes large the distribution tends to Poisson.

(Hint: $\lim_{N\to\infty}(1+x/N)^N=\exp x$)

5.4 In (5.3.5) it is supposed that the two different electron populations are kept in equilibrium with the photons and that the quasi-Fermi levels for the upper and lower levels are $\mathscr{E}_{fn}$ and $\mathscr{E}_{fp}$, respectively.[55] Show that, with no escaping photons, the radiation density is

$$P=(A/B)/\{\exp[(hf-\mu)/kT]-1\}$$

where $\mu = \mathscr{E}_{\mathrm{fn}} - \mathscr{E}_{\mathrm{fp}} < hf$. If $\mu > hf$, explain qualitatively what would happen.

5.5 Suppose that there are n_i electrons (fermions) at $\mathscr{E}_i$, where there are N_i states ($i = 1$ or 2) which interact with c_r classical systems in C_r classical states, which lie in energies $\mathscr{E}_r$ ($r = 3$ or 4), where $\mathscr{E}_1 - \mathscr{E}_2 = \mathscr{E}_3 - \mathscr{E}_4$. Explain the rate equation

$$\mathrm{d}n_1/\mathrm{d}t = An_1C_3(N_2 - n_2)c_4 - An_2C_4(N_1 - n_1)c_3$$

Given that in equilibrium $c_r/C_r = \exp(-\mathscr{E}_r/kT')$, and

$$n_i/N_i = 1/\{1 + \exp[(\mathscr{E}_i - \mathscr{E}_f)/kT]\}$$

show that $kT = kT'$.

5.6 Consider a communications system where, over a period of time, 10^9 pulses are sent out. On average, each pulse contains 20 photons. The system is sensitive enough that even if 1 photon can be detected the pulse is registered, so that errors arise only through the possibility of zero detection of photons in the pulse in spite of a mean number of photons of 20. How many errors (i.e. failure to record a pulse) are expected over the 10^9 pulses for (a) Poisson distribution, (b) single mode chaotic light distribution?

Rate equations in optoelectronic devices

6.1 Models for optoelectronic devices

6.1.1 *Introduction*

The work here concentrates on semiconductor devices, leaving gas and atomic lasers for further reading.[56–59] This chapter considers tutorial models for the diode injection laser, the semiconductor light emitting diode (LED) and photodiode (PD), all of which can be understood from the same forms of rate equations, though with different constraints and interpretation. It will be helpful to have models for these three devices before starting on their rate equations.

6.1.2 *The light emitting diode*

The LED is a p–n junction driven by a current I into forward bias. Electrons and holes recombine close to the junction between the p- and n-materials and give out radiation with a frequency determined by the photon energy which in turn is determined by the material's impurities or band gap.[60] Some fraction of the forward current I is then turned into useful light L (power) formed from photons with a mean energy hf_m. The quantum efficiency is $\eta = (eL/hf_m I)$. The more efficient LEDs have a well-defined recombination region (volume Φ). For example, in GaAs LEDs (Fig. 6.1), the region Φ can be defined through the use of 'heterojunctions', where a p-type GaAs layer is sandwiched between n- and p-type $Ga_{1-x}Al_xAs$ materials which have a wider energy gap between conduction and valence band than GaAs materials. On forward bias, holes are driven from the p-$Ga_{1-x}Al_xAs$ material into the GaAs, but the potential difference that arises between the valence band of p-type GaAs and n-$Ga_{1-x}Al_xAs$ prevents the holes diffusing into the n-$Ga_{1-x}Al_xAs$ material. (Typically with $x \sim 0.25$, the potential barrier is $\sim$0.3 eV.) Similarly electrons driven from the n-side, into the p-GaAs, cannot diffuse into the p-$Ga_{1-x}Al_xAs$ because of a potential barrier in the conduction band. The p-GaAs layer (acceptor density N_A) then defines the region Φ in which the intense recombination occurs.

The light in this region has a photon density P with a spread of frequencies δf ($h\delta f \sim 2kT$). The useful light output is caused by the photons escaping at a rate which is given from P/τ_p, tending to reduce the photon density. Of course, the generation of light must balance this loss, at least in the steady state. The evaluation of this rate of loss is discussed in section 6.2.4, and the time scale is typically in the picosecond to femtosecond range.

6.1.3 *Semiconductor diode lasers*

Laser injection diodes (Fig. 6.2) and light emitting diodes[8,14,23,60–64] often use similar construction techniques with suitable materials such as GaAs/$Ga_{1-x}Al_xAs$ or quaternary compounds using In, Ga, Al, As, P. For diode lasers, heterojunction technology has been vital as a technique to confine the electrons and holes into well-defined regions for strong stimulated recombination. Important additional features of the laser are the reflecting facets at the ends of the crystal. These facets form a resonant optical cavity (a Fabry–Perot resonator) which increases

Fig. 6.1. A schematic design for a light emitting diode. (*a*) Structure (shaded zone has been proton bombarded to make it semi-insulating – see section 2.3). (*b*) Electron energy ($\mathscr{E}$) band – distance diagram. Barrier $\Delta\mathscr{E}$ inhibits electrons and holes from moving into higher band gap material over these potential barriers. Recombination confined to p-GaAs region, volume Φ.

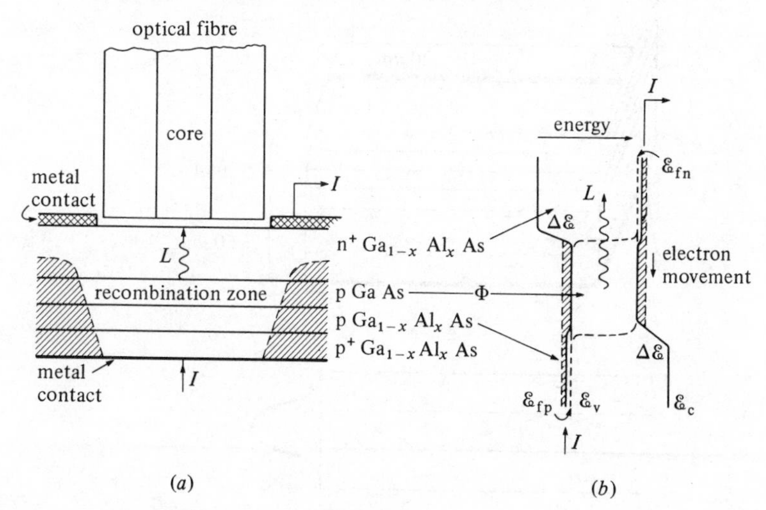

the optical intensity inside the laser for a given power output, selects stimulated emission in primarily one direction, and ideally selects a single optical mode which has the strongest interaction with the electrons and holes in the semiconductor.

One of the key features of stimulated emission is that the stimulated photons have the same frequency and direction as the stimulating photons and increase the electromagnetic wave's amplitude without altering the phase. Spontaneous emission, although it has the same frequency, is effectively emitted in an arbitrary direction and does not necessarily maintain the phase of any existing electromagnetic wave. Spontaneous emission is essentially like noise over a limited frequency range.

Fig. 6.2. A schematic design for a semiconductor injection laser. (*a*) Structure: metal contacts (M) taking current I to interaction region Φ formed from GaAs (layer 3) sandwiched between $Ga_{1-x}Al_xAs$ p-type (layer 2) and $Ga_{1-x}Al_xAs$ n-type (layer 4). Layers 1 and 5 are GaAs layers chosen to make better low resistance contacts. Note scales are distorted. (*b*) Waveguiding mechanisms: GaAs layer 3 is of slightly higher refractive index than GaAlAs outer layers 2 and 4. This leads to total internal reflection for light at low enough angles of incidence giving a guiding mechanism vertically. Horizontal guiding is achieved by change of complex refractive index (gain guiding) reflecting wave back and forth between region with dense electrons and region with few electrons. Facets at either end reflect as well as emit light and provide an optical resonator.

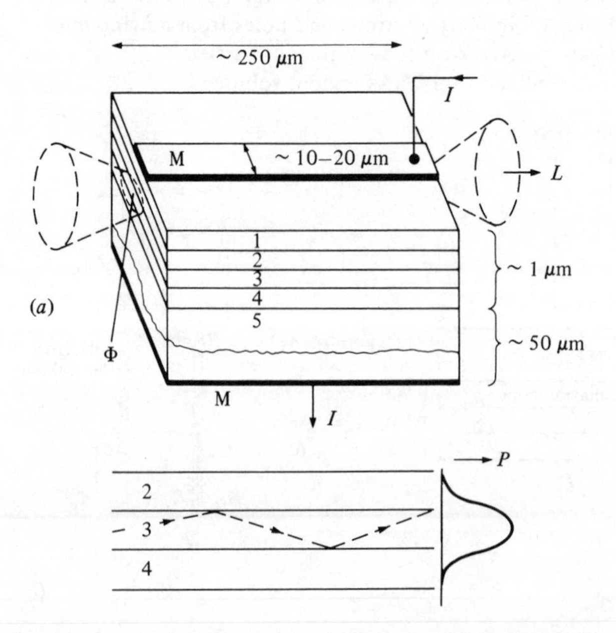

In a laser, the photon density P is higher than in an LED and is formed from photons of nearly a single photon energy hf. There is also usually a waveguide built into the device to keep the photon intensity high and close to the recombination region of the electrons and holes.

In the structure shown, one of the waveguiding mechanisms is fortunately given by the same heterojunction technology which helps to confine the electrons and holes into a local recombination region. The p-GaAs layer has a slightly higher refractive index than the surrounding material (p- or n-type $Ga_{1-x}Al_xAs$) so that light is guided along this layer similar to the way that light can be guided along a glass fibre in air.[23,64] This surprising 'hat-trick' of confining photons, electrons and holes using a single technology has been a key feature in the success of making injection lasers capable of operating continuously at room temperature. The use of quaternary materials has extended this hat-trick of effects by permitting the band gap to be engineered by choice of the material composition. The lasing wavelength can currently be controlled over a range 1.3 to 1.6 μm, of value in telecommunications using optical fibres.

6.1.4 *The photodiode*

The model for a simple photodiode is again a p–n junction. It has highly doped p^+- and n^+-contact regions, and most of the material between these p^+- and n^+-contacts is taken to be intrinsic: a p–i–n diode (Fig. 6.3). This i-region forms the interaction region of volume Φ and

Fig. 6.3. A schematic cross-section of a photodiode (p–i–n). Wide band gap materials for contacts prevent absorption in contacts. Photons absorbed in ν-region (high purity material ideally 'intrinsic') producing hole–electron pairs. Device reverse biased. If heterojunction technology not possible then contact layers must be thin to avoid absorbing light and slowing down response. (Proton bombardment gives the material the property of high resistivity.)

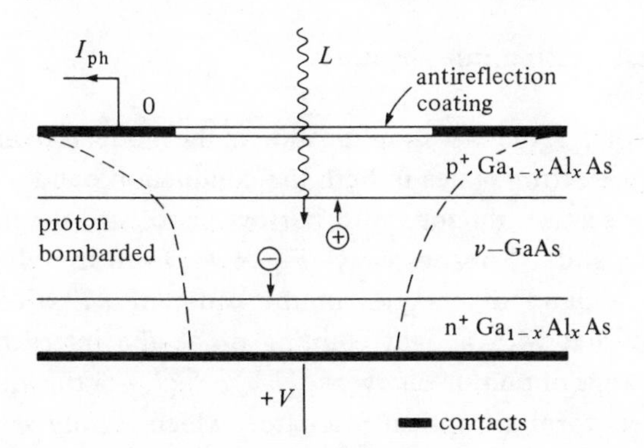

with $N_D \sim 0$. Under normal operation, the diode is reverse biased so that the i-region is depleted of charge carriers except for photogenerated holes and electrons. The electric field within the i-region forces such photogenerated holes and electrons to travel at their scattering limited velocities towards the p- and n-regions, respectively. The i-region also keeps the device's capacitance low so that rapid changes in the photogenerated current are not shorted out by a large capacitance. The composition of this i-layer is chosen so that the band gap energy is less than the incoming photon energy, and the length of the i-region is chosen to absorb the photons effectively. For greatest efficiency and speed, the p- and n-contacts should absorb negligible light and store negligible numbers of minority charge carriers. This can be engineered by making the contacts very thin or by making them from a different material with a band gap higher than the photon energy so that the incoming photons cannot be absorbed effectively except in the i-region.

To highlight the key physics of all these devices, the electron, hole and photon densities will be assumed initially to be average uniform values. A filling factor can be used to account for the fact that the electrons and photons may be confined to slightly different regions of volume Φ and Φ'. The approximations in all these as assumptions are not too severe and are certainly useful for tutorial models. For example, although there is interference of the optical fields within the laser cavity, giving submicron variations of intensity and interaction, these can be treated as an average field, especially because diffusion of the electrons smooths out the spatial rates of change of electron density over the scale of a half-micron or so. In more detailed models it is found that within the cavity there can be a variation around $\pm 10\%$ in the average optical intensity because the average photon density increases towards the facets.

We can turn then to the rate equations for the photons in such a general interaction region.

6.2 Photon and electron rate equations

6.2.1 *Introduction*

As in section 5.3, the first simplification in the model is to assume that the interacting electron states in both the conduction band and the valence band of a semiconductor lie in narrow bands of energies $\Delta\mathscr{E}$ wide centred at $\mathscr{E}'_c$ and $\mathscr{E}'_v$, respectively (Fig. 6.4). Typically electrons and holes fill up a band of energies of the order of kT wide in a semiconductor, so that with a light emitting diode the interaction is concerned with a range of photon energies $\Delta(\hbar\omega) \sim 2kT$. For the injection laser, cleaved facets form an optical resonator, which ideally selects a single mode where the photon–electron interaction is strongest within this range of energies. In this ideal case only a single optical mode with the single photon energy $h\nu = \mathscr{E}'_c - \mathscr{E}'_v$ need be considered. Typically $\mathscr{E}'_c$ and $\mathscr{E}'_v$ are found to be about kT units of energy higher and lower than the accepted respective values of band edge energies $\mathscr{E}_c$ and $\mathscr{E}_v$.

For the photodiode, one need only consider one wavelength of energy coming in at a time. The device is found to be sufficiently linear that the effect of a combination of wavelengths can be discovered by superposition. Thus for this device only a narrow range $\Delta(\hbar\omega)$ need be considered.

6.2.2 *Validity of model*

We digress for the moment to the utility and validity of this model for a semiconductor. A reader may skip over this section at a first reading.

Fortunately, in devices considered here, the electrons move between adjacent levels of the conduction band at extraordinarily high rates, so that, if one energy level becomes relatively too full or depleted compared to neighbouring energies, then 'thermalisation' by movement of the electrons into or from neighbouring energy states within the conduction band can occur on picosecond time scales through interactions between the electrons themselves or with the lattice. Similar remarks apply to holes in the valence band. Simple rate equations using a single effective energy level for the conduction and for the valence band have certainly proved to be helpful in understanding the performance of semiconductor injection lasers down to time scales of tens of picoseconds. On much shorter time scales, it may be expected that there is local structure of the

Fig. 6.4. A schematic diagram for photon–electron interactions in a semiconductor device. $\mathscr{E}'_c$ and $\mathscr{E}'_v$ represent effective energies for the conduction and valence bands with effective density of states N_c and N_v containing n electrons and p holes. Device taken to be uniform over interaction volume Φ. τ_p = time scale for escape of photons from interaction volume; τ_r = time scale for recombination.

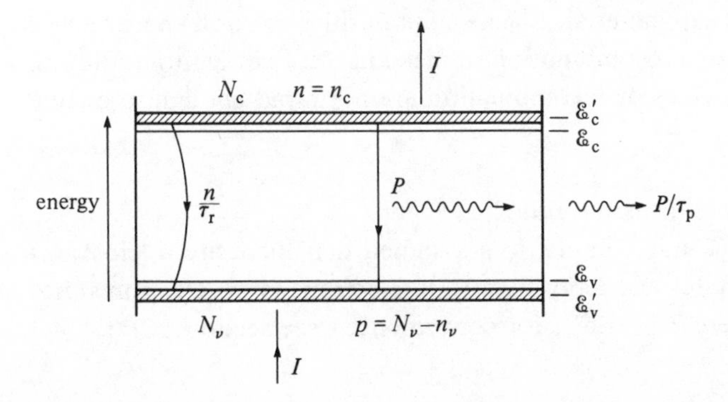

occupation of energy states by electrons, because the photons too rapidly deplete one energy level ('hole burning') or one region of space ('spatial hole burning'). Further equations considering the rates of redistribution of electrons and holes may have to be considered, but this is not done here.

Another important caveat is concerned with conservation of momentum when electrons, holes and photons interact as introduced briefly in Fig. 2.6. Typically at present, injection lasers can be made with specific wavelengths λ in the micron range. The photons have a momentum h/λ, while thermal electrons have momentum $\sim h/L_t$ (measured in the same units) with L_t of order 10 nm at room temperature (taking $m^* = m$). Semiconductor crystals have their scale of spatial periodicity determined from their lattice parameter a, which has values typically around 5 nm. This periodicity determines the maximum magnitude of momentum ($\sim h/2a$) for an electron in the crystal. Any momentum in excess of this value is transferred from the electron to the crystal as a whole. Because $\lambda \gg L_t$, photons have an almost negligible amount of momentum compared to electrons (or holes) in the crystal. To conserve overall momentum in the exchange between photons, electrons and holes, the electrons must fall in energy with a relatively small change of momentum.

In a semiconductor, this limits good radiative interaction between photons, electrons and holes to 'direct band gap' materials. Engineering of direct band gap compound semiconductors has leapt forward with improved methods of growing materials such as $Ga_xAl_{1-x}As$ ($x < 0.35$), and indeed numerous quaternary compounds of In, Ga, Al, As, P. In such direct gap materials, the conduction band energy-momentum minimum is directly above the maximum in the valence band (Fig. 2.5). Silicon and germanium are indirect gap semiconductors (Fig. 2.6), and band to band recombination requires the addition of momentum of the order of $h/2a$ which can come from the lattice vibrations (phonons). However, the simultaneous confluence of photons, phonons, electrons and holes of the right energy and momentum is a relatively improbable process so that radiative recombination will occur at a lower rate than in a direct band gap material. Localised impurities can add energy levels which aid radiative recombination to make useful light emitting diodes, as in Si, but the rates of recombination are not rapid enough to enable laser action to occur.

6.2.3 *Stimulated emission rates*

Taking a unit volume, it is assumed that there are n_c electrons in N_c electron states at energy $\mathscr{E}_c$ with n_v electrons in N_v electron states at energy $\mathscr{E}_v$ (here $\mathscr{E}'_{c/v} \rightarrow \mathscr{E}_{c/v}$ for convenience – see section 6.2.1).

Following then the discussion of net stimulated emission as the second term in (5.3.5) we arrive at

$$\mathrm{d}P/\mathrm{d}t)_{\text{net stimulated}} = BP[n_c(N_v - n_v) - n_v(N_c - n_c)] \quad (6.2.1)$$

where the spontaneous component has been omitted for the present. The reader should check through this discussion with the usual arguments about available electrons for transfer and the available sites into which the transfer can occur.

In terms of conduction electron density n and valence band hole density p

$$n = n_c, \quad p = N_v - n_v \quad (6.2.2)$$

At this stage it is helpful to assume that the electrons and holes are so dense that they have to adjust to approximate electrical neutrality within the region where the recombination is occurring. It was seen (section 2.2) that a dense enough mobile set of electrons with a positive background could adjust to become electrically neutral in subpicosecond time scales, so that a further rate equation for this adjustment can be omitted provided we work on considerably longer time scales than the subpicosecond dielectric relaxation time (see also problems 1.5, 1.6).

In some current laser technology the recombination region is a highly doped p-type region with N_A ionised acceptors, so that electrical neutrality implies

$$p = N_A + n \quad (6.2.3)$$

(6.2.1), (6.2.2) and (6.2.3) lead to

$$\mathrm{d}P/\mathrm{d}t)_{\text{net stimulated}} = GP(n - n_o) \quad (6.2.4)$$

where

$$G = B(N_c + N_v); \quad n_o = N_c(N_v - N_A)/(N_c + N_v) > 0$$

The value of G is the gain constant for the material and n_o is the density of the injected electrons at which the material appears transparent; i.e. there is no net stimulated absorption or recombination. Care must be taken in interpreting this equation. With high doping, N_c and N_v will not be exactly the same effective density of states as at low doping (see standard calculations, references 5, pp. 151–4; 8, pp. 17, 18; 9, p. 141), but will be expected to take similar orders of magnitude around 10^{24}–$10^{25}/\mathrm{m}^3$. It must still be expected that the ionised acceptor density N_A is less than the effective density of states N_v so that n_o has a fundamentally positive value. The gain G, a function of electron density and wavelength, is quite difficult to calculate accurately,[62,63] and it is taken here as experimentally determined: $G \sim 10^{-12}$ m^3 s as a rough guide giving growth time constants around 1 ps for excess densities around $10^{24}/\mathrm{m}^3$.

Often the optical field extends outside the interaction region Φ as in Fig. 6.2 so that the number of photons actually interacting with electrons in the region Φ is reduced by some fraction F calculated from the field profiles. Authors who make use of this *confinement* factor F write

$$\mathrm{d}P/\mathrm{d}t)_{\text{stimulated}} = GFP(n - n_o) \tag{6.2.5}$$

In most of the work here, the electron and photon densities are assumed to be averaged over the same volume Φ with appropriate scaling. The additional losses of photons escaping sideways from the interaction region can be accounted for separately (see section 6.2.5).

6.2.4 *Spontaneous emission rates and coupling*

Added to the net stimulated emission is the spontaneous recombination. If the discussion of chapter 2 is followed, then the recombination is approximately determined by a single time constant τ_r. However, recombination takes two forms: radiative recombination and non-radiative recombination. It is better then to write

$$\mathrm{d}n/\mathrm{d}t)_{\text{spontaneous}} = -(n - n_e)/\tau_r \tag{6.2.6}$$

where

$$1/\tau_r = (1/\tau_{rr}) + (1/\tau_{nr})$$

By definition here the spontaneous radiative recombination gives out light at the same optical wavelengths as the photons represented by the density P. The 'non-radiative' recombination may well give out photons, as well as requiring phonons, but any photons are at different frequencies than the principle photons (P) under discussion.

Even when the emission is at the correct frequency, not all the spontaneous emission from the radiative recombination will couple into the laser's waveguiding mechanisms. For a tutorial model, a spontaneously emitted photon is radiated in any one direction with equal relative

Fig. 6.5. Diagram to illustrate spontaneous coupling factor. (a) Spontaneously emitted photon has probability of $\delta\sigma/4\pi$ of coming off from area $\delta\sigma$ in direction P. (b) Coupling to waveguide for spontaneous event at centre $\sim 2A/(\pi L^2)$.

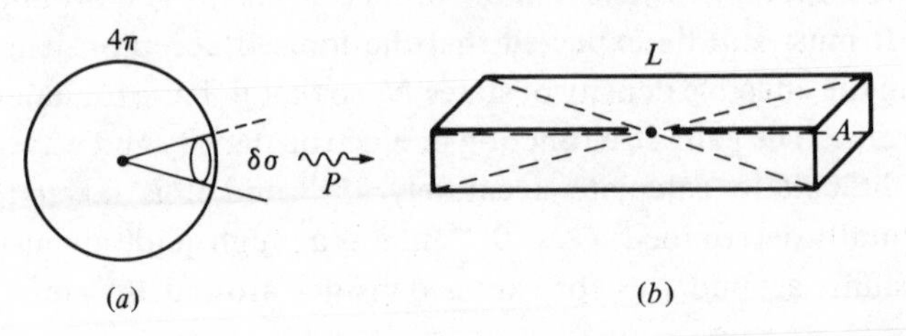

probability of $(1/4\pi)$ per steradian (Fig. 6.5). (In a more sophisticated analysis, the spontaneous emission excites all optical modes with equal probability.) An injection laser with a waveguide cross-section of A (typically here around 1 to 10 μm^2) and length L then represents a solid angle to spontaneous emission in the centre of the laser of approximately $2(4A/L^2)$. The factor of 2 indicates that the spontaneous light can couple to stimulated light travelling in either direction along the length of the laser. The probability of a spontaneous photon coupling into the stimulated photons is then given by $\beta \sim 2A/(\pi L^2)$. For $L \sim 200$ μm, a spontaneous coupling factor β lower than 10^{-3} is readily achieved.

The details of the optical waveguiding can alter this coupling factor. Stripe lasers, with no significant changes of refractive index in the direction parallel to the p–n junction interface, are referred to as 'gain guided' because the stimulated emission and gain determine how the optical field spreads transversely. Such guiding produces a relatively large curvature to the optical wavefronts propagating through the laser, thereby giving a wider range of solid angles for spontaneous emission coupling compared with lasers with built in optical waveguides. In lasers with strong optical guides, the wavefronts are relatively straight, giving a minimum range of solid angles for coupling to spontaneous emission.

The rate of spontaneous emission (neglecting the equilibrium density $n_e \ll n$) into the photon density P is then given from the radiative recombination by the rate equation

$$\mathrm{d}P/\mathrm{d}t)_{\text{spontaneous}} = \beta n/\tau_{rr} \tag{6.2.7}$$

6.2.5 *Photon lifetime*

The rate of loss of photons into external regions, where they potentially can be used, is given by P/τ_{pe}, while equally photons are driven from any external density P_e into the interaction region at the rate P_e/τ_{pe}. These rates are determined by the mechanisms which determine how photons escape from the interaction region.

In a typical semiconductor injection laser the facets may have dielectric coatings to change the power reflectivity to values R_1 and R_2. Without any coating, the theory of electromagnetic waves shows that there is a power reflection given from $R = [(1-\mu_r)/(1+\mu_r)]^2$, where μ_r is the refractive index of the material (for material with no magnetic effects $\mu_r = \varepsilon^{1/2}$, with ε the relative permittivity). R is then typically close to 0.33 for uncoated GaAs lasers and takes a slightly higher value for the longer wavelength injection lasers using InP based materials.

The photon lifetime can be estimated by assuming a uniform electron density n_t giving a gain $G_t = G(n_t - n_o)$ for the stimulated emission within the laser. The photon density then grows as

$$P = P_o \exp 2\alpha z$$

where

$$2\alpha = G_t / c_{gm} \qquad (6.2.8)$$

and c_{gm} is the velocity for the photon energy propagating from facet 1 and facet 2 and back. For a steady state of emission, the photons reflected back to the start in a single whole round trip (time $\tau = 2L/c_{gm}$) must have the same density. Now if P_o photons start out from facet 1, $R_2 P_o \exp(G_t \tau/2)$ are reflected from facet 2 to grow with another pass down the laser. Then $R_1 R_2 P_o \exp(G_t \tau)$ are reflected from facet 1 to start again. Hence, for a steady state

$$(R_1 R_2 P_o) \exp G_t \tau = P_o$$

or on rearranging

$$G_t = [c_{gm} \ln(1/R_1 R_2)]/2L \qquad (6.2.9)$$

In terms of the steady state averaged density rate equations, ignoring spontaneous emission:

$$\mathrm{d}P/\mathrm{d}t = [G_t - 1/\tau_p]P = 0 \qquad (6.2.10)$$

giving

$$G_t \tau_p = 1$$

or

$$\tau_p = 2L/[c_{gm} \ln(1/R_1 R_2)] \qquad (6.2.11)$$

For $L = 100$ μm the photon lifetime is typically a little under 1 ps.

The effect on the overall photon lifetime of photon loss within the cavity needs some extra discussion. Both stimulated emission and stimulated absorption have been included to give a net emission rate that is positive for electron densities in excess of a transparency value (see (6.2.4)). However, often tails of the light distribution can extend sideways (Fig. 6.2(b)) outside the real interaction region so that there is some additional absorption not fully included in the term $-Gn_oP$. There is an additional loss rate $-P/\tau_{pi}$ which has to allow for this absorption of poorly confined light. The overall photon lifetime is then

$$1/\tau_p = (1/\tau_{pe}) + (1/\tau_{pi}) \qquad (6.2.12)$$

where $1/\tau_{pe} = [c_{gm} \ln(1/R_1 R_2)]/2L$ gives the external losses and $1/\tau_{pi}$ gives internal losses.

For light emitting diodes, the photon lifetime will be similarly defined from the rate at which photons can escape from the interaction region. However, without the feedback of photons from reflecting facets, the photon density is sufficiently low that stimulated emission is not as important as spontaneous emission in the LED.

6.2.6 *Net photon rate equations*

Combining the stimulated emission, spontaneous emission and the various losses of the photons gives the total rate for the increase of photons as:

$$\mathrm{d}P/\mathrm{d}t = GP(n-n_o)+(\beta n/\tau_{rr}) - (P/\tau_{pi})-(P/\tau_{pe})+(P_e/\tau_{pe}) \tag{6.2.13}$$

To summarise, the first term on the r.h.s. is the net stimulated emission in the interaction region. The second term is the spontaneous emission, which couples into the same wavelengths and modes as the P photons being considered in the rate equation. For lasers this coupling factor β can be low (around 10^{-3}). The third term gives the rate of absorption of photons caused by any optical fields extending outside the electron gain region. The fourth term gives the light output, i.e. the rate of supply of potentially useful photons which escape from the ends of the interaction region (subscript 'e' for loss of photons externally contrasting with subscript 'i' for internal loss). The final term gives the rate of increase of photons from any external supply.

6.2.7 *Travelling wave rate equations*

While (6.2.13) is satisfactory for 'lumped' interactions where the device is considered as a whole, it needs to be modified for those interactions where the photons are travelling forward into a material, as in photodiodes or non-uniform lasers. For travelling waves then we note the photon energy flux is $c_{gm}P$, where c_{gm} is the group velocity at which energy propagates in the material. Returning to the continuity relationships discussed in section 1.2, it follows that

$$(\partial P/\partial t)+(1/c_{gm})(\partial P/\partial z) = (\text{net rate of gain of photons}) \tag{6.2.14}$$

Thus for travelling waves of photons

$$(\partial P/\partial t)+(1/c_{gm})(\partial P/\partial z) = GP(n-n_o)+(\beta n/\tau_{rr})-(P/\tau_{pi}) \tag{6.2.15}$$

The external loss is subsequently accounted for by appropriate spatial boundary conditions for the photon density at the facets of the laser.

6.2.8 *Electron rate equations – the drive*

The interaction volume Φ contains a density of n electrons which are recombining with a density of p holes. With the assumption of electrical neutrality $dn/dt = dp/dt$, so that only the electrons need be considered.

The spontaneous recombination removes electrons (see (6.2.6)) at a rate $[(n - n_e)/\tau_r]$, where n_e is the equilibrium density. Usually for light emitting diodes or lasers $n \gg n_e$, so n_e will be neglected here. There is no need here to worry about which, if any, photons are emitted because all types of recombination remove the electrons from participating in any further interaction. Added to this spontaneous recombination is the stimulated recombination of electrons, given by (6.2.4), at a rate $PG(n - n_o)$.

The current I, driving into the volume Φ, gives an increase in the density n at a rate of $I/(q\Phi)$. These rates of decrease and increase, with appropriate signs, give for a unit volume

$$dn/dt = (I/q\Phi) - (n/\tau_{rr}) - (n/\tau_{nr}) - PG(n - n_o) \tag{6.2.16}$$

An optical input of photons, at energies higher than the band gap energy and not at the same photon energy as the photons of density P, could form another drive term contributing to dn/dt. This further complication with different photon energies is not considered here.

6.2.9 *Travelling wave electron drive*

For completeness, the travelling wave rate equations for electrons are briefly discussed. From the continuity of current

$$(\partial n/\partial t) + [\partial(vn)/\partial z] \\ = -(n/\tau_r) - D(\partial^2 n/\partial z^2) - GP(n - n_o) + J \tag{6.2.17}$$

where the r.h. terms give the net rate of loss. The diffusion current $-D(\partial^2 n/\partial z^2)$ removes electrons to adjacent sections. In a full model there will also be transverse diffusion not included in (6.2.17). Over a finite enclosed volume $\iiint D \operatorname{div}(\operatorname{grad} n)\, dV = \iint D (\operatorname{grad} n) \cdot dS = 0$ given that the diffusion currents $-D \operatorname{grad} n$ vanish at the boundaries. In many optical devices, there is negligible mean velocity for the carriers, so $v \sim 0$ and

$$(\partial n/\partial t) + (D\partial^2 n/\partial z^2) + (n/\tau_r) = -GP(n - n_o) + J \tag{6.2.18}$$

If P varies, for example, as $(1 + m \cos 2\beta z)$ because of *standing* electromagnetic waves inside the device then variations are induced in n. However, such ripples in n at $\cos 2\beta z$ decrease with β and $D\tau_r$ as $1/(4\beta^2 D\tau_r)$ and, with large enough $D\tau_r$, can be neglected so that n is approximated by a uniform distribution.

6.3 Rate equations for the photodiode

6.3.1 *Optical absorption*

The scene is set by considering a steady input of light into an interaction region with negligible free carriers ($n = 0$) and rewriting the travelling wave rate equation (6.2.15) as

$$dP/dz = -\alpha P \tag{6.3.1}$$

where $\alpha = G n_o c_{gm}$. The time derivative is ignored for steady light. Loss outside the interaction region has also been ignored for the present. The photons then are absorbed as $P = P_o \exp(-\alpha_o z)$, and α_o is the optical absorption coefficient for the material and the particular wavelength (Fig. 6.6). Light is absorbed strongly by the photons stimulating hole–electron generation once the photon energy exceeds the band gap energy. A variety of effects can absorb the photons at other energies,[14] but these are usually weak compared with the band gap absorption. The simple rate equation model predicts the exponential absorption, but the required detailed calculations to determine α as a function of wavelength cannot be done here.

6.3.2 *A 'lumped' model for the photodiode*

The photogeneration of hole–electron pairs is proportional to the light and so the photogenerated electronic charge decays with depth in the photodiode as the light is absorbed. However, the total electronic

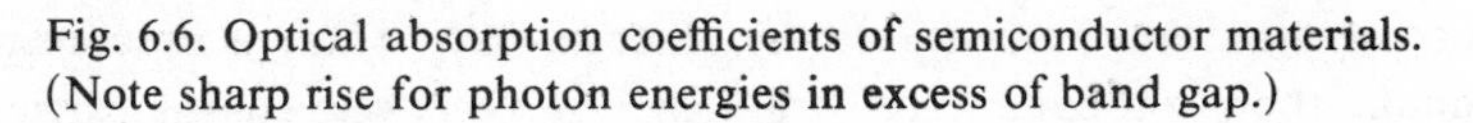

Fig. 6.6. Optical absorption coefficients of semiconductor materials. (Note sharp rise for photon energies in excess of band gap.)

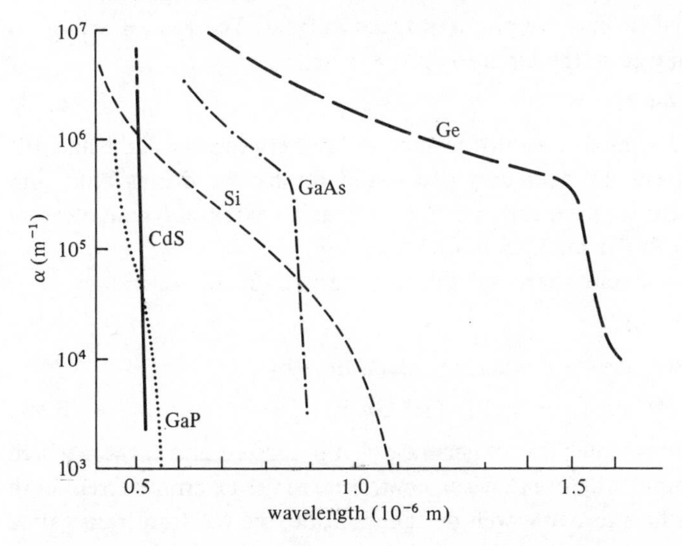

charge Q that is photogenerated will define an average electron density $n = Q/q\Phi$. Similarly an average photon density P may be defined over the interaction volume Φ for a p–i–n diode (Fig. 6.3). In normal operation the diode is reverse biased so that the charge carriers are swept out at a rate n/τ, where τ is the charge carriers' effective transit time which is taken to be limited by scattering (~10 ps/μm). It is assumed that no charge carriers are generated or stored in the p- and n-contacts. The electron density n will be assumed to be negligible compared to the transparency density n_o at which stimulated emission balances stimulated absorption. From (6.2.13), the photon rate equation then becomes approximately

$$dP/dt = -GPn_o - (P/\tau_{pi}) + (L/\Phi) \tag{6.3.2}$$

where $L = (\Phi P_e/\tau_{pe})$ photons/s is the light input from the external photon field P_e into the volume Φ. The rate of loss or absorption which does not generate charge carriers is given by P/τ_{pi}.

There is no current drive into the device but the electron rate equation (6.2.16) must be modified to allow for the reverse bias fields removing the charge carriers at a first-order rate of n/τ, where τ is the effective transit time for the carriers, and at a rate n/τ_r for recombination:

$$dn/dt = PGn_o - (n/\tau_r) - (n/\tau) \tag{6.3.3}$$

Each photon induces one hole–electron pair with the electrons driving current into one contact and the holes preserving current continuity by driving the same current into the other contact. There is no current gain unlike the photoconductor (problem 3.6). Only one charge carrier need be considered to find the photo-induced current. The rate of arrival of electronic charge to the circuit is given from

$$I = q\Phi n/\tau \tag{6.3.4}$$

where it is assumed that all the time delay between the light and the external current is adequately accounted for by the delays built into (6.3.3). The device's capacitance is treated as an external circuit element in parallel with the load, as in (3.2.9).

In the steady state there is a linear current/light characteristic:

$$I/q = \eta L$$

where η is the quantum efficiency given here by

$$\eta = [(Gn_o\tau_{pi})(\tau + \tau_r)]/[\tau_r(1 + Gn_o\tau_{pi})] \tag{6.3.5}$$

Clearly photons which do not generate charge carriers and carriers which recombine rapidly before they can contribute to the external current both reduce the efficiency. In a well-designed diode, the electron transit time is much shorter than the recombination time, and (6.3.3) may be recast approximately as

$$\tau'(\mathrm{d}I/\mathrm{d}t)+I=\eta qL \tag{6.3.6}$$

where $\tau'=\tau_r\tau/(\tau_r+\tau)$. Light absorbed within the contact layers changes this response (see problem 6.5). The photo-induced current I drives into the device's capacitance and load (along with contact resistances and lead inductances), and all must be considered in a more extensive discussion on the overall speed within a circuit.

6.3.3 *The avalanche photodiode (APD)*

Avalanche multiplication can be used in photodiodes to enhance the current for a given photo-input. Impact ionisation was discussed briefly in section 3.2. The effect can be dealt with in a phenomenological way by saying that, at high electric reverse bias fields, one photogenerated hole–electron pair has a probability of producing $\int_\Phi \alpha_i\,\mathrm{d}x$ more pairs by impact ionisation before reaching the contacts to the photodiode. The ionisation rate per unit distance is α_i and varies rapidly with reverse bias electric field E_r. A useful guide is $\alpha_i\sim(1/L_i)\exp(-E_c/E_r)$, where E_c is some characteristic field and L_i a characteristic length. (For electrons in Si: $E_c\sim3.5\times10^7$; $L_i\sim90$ nm.) For brevity this discussion neglects differences between holes and electrons for their rates of ionisation.[8,11,23]

(6.3.6) is then recast in terms of the rate of change of charge Q of one of the charge carriers with the external current $I=Q/\tau$:

$$\begin{aligned}\mathrm{d}Q/\mathrm{d}t)_{\text{loss to contacts}}&=-Q/\tau\\ \mathrm{d}Q/\mathrm{d}t)_{\text{gain from impact ionisation}}&=Q\int\alpha_i\,\mathrm{d}x/\tau\\ \mathrm{d}Q/\mathrm{d}t)_{\text{gain from light}}&=\eta qL\end{aligned} \tag{6.3.7}$$

Neglecting recombination, (6.3.6) becomes

$$\tau\,\mathrm{d}I/\mathrm{d}t+I\left(1-\int\alpha_i\,\mathrm{d}x\right)=\eta qL \tag{6.3.8}$$

In the steady state

$$I=M\eta qL \tag{6.3.9}$$

where the multiplication factor is

$$M=1\bigg/\left[1-\int\alpha_i\,\mathrm{d}x\right]\sim1/[1-(V_r/V_b)^m] \tag{6.3.10}$$

The empirical formula on the right has V_r as the reverse bias voltage, V_b the 'breakdown' voltage and m of order 6 or 7. The value of M is infinite when $\int\alpha_i\,\mathrm{d}x=1$ or $V_r=V_b$. The device breaks down at this voltage as any initial charge gives 'infinite' current. Practical multiplications of $M \sim 100$ to 1000 are the normal range of operation, but the reverse bias voltage has to be critically controlled to keep the multiplication stable.

The dynamic response is considerably slowed down[8,11,23] by large values of M because the time constant in (6.3.8) is $M\tau$. This rate equation demonstrates then the necessity of having very short regions for avalanche multiplication if avalanche gain in current is not to be simply lost in a slower speed of the device. Clever construction techniques are used to effectively give narrow avalanche regions with picosecond transit times (Fig. 6.7) so that multiplication is gained with good optical absorption and good response times. Low absorption at the contacts is again required.

6.4 Injection laser dynamics

6.4.1 *Threshold conditions*

The rate equations of (6.2.13) and (6.2.16) can be applied directly to an injection laser where a single optical mode, photon density P, is

Fig. 6.7. APD construction. (*a*) Schematic cross-section for a possible Si device. (*b*) Field profile from n^{++}-p^{+}-π-n^{+} 'Read type' diode[11] device on reverse bias. p-region produces narrow peak to field so avalanche zone <1 μm thick. Avalanche initiated by electrons moving from absorption zone to n^{++}-contact. Electrons have higher ionisation rates than holes so electron initiated ionisation gives required current for fewer collisions and fewer fluctuations of collisions, hence giving less noise.

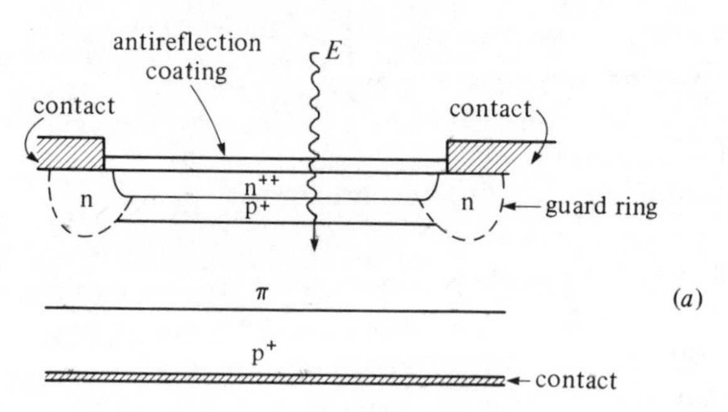

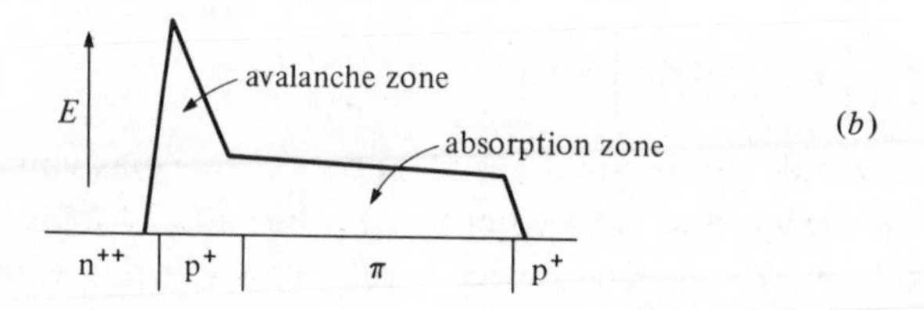

excited. The required features may be demonstrated from

$$dP/dt = GP(n - n_o) + (\beta n/\tau_r) - (P/\tau_p) \tag{6.4.1}$$

Recapitulating, G gives the gain, with n_o an effective transparency density assuming no excess internal loss outside the interaction region. The second term gives the spontaneous emission coupling to the lasing mode. The final term gives the potentially useful photons.

The electron rate equation is given from (6.2.16)

$$dn/dt = -(n/\tau_r) - GP(n - n_o) + D \tag{6.4.2}$$

where $D = I/e\Phi$ is the drive.

The results are highlighted by assuming $\beta = 0$ initially so that in the steady state with $d/dt = 0$, (6.4.1) requires either

$$P = 0 \quad \text{or} \quad n = n_t = n_o + 1/(G\tau_p) \tag{6.4.3}$$

The value n_t is referred to as the threshold density. If $P = 0$, then (6.4.2) yields

$$n/\tau_r = D$$

If $P \neq 0$, then

$$GP(n_t - n_o) = (D - D_t) = P/\tau_p$$

where

$$D_t = n_t/\tau_r$$

The potentially useful light output from one of the two facets is then

$$L = P\Phi/2\tau_p = \tfrac{1}{2}(I - I_t)/q \tag{6.4.4}$$

where $I = qD\Phi$ and the light output is measured in photons/s.

In this ideal model all the injected current above a threshold value is turned into light so that the single facet optical efficiency is $\sim(I - I_t)/2I$. Additional internal losses for photons outside the interaction laser will reduce this efficiency (see problem 6.4).

The ideal light/current characteristics are shown in Fig. 6.8. There is no light output until the threshold current I_t is reached with $n = n_t$, and after that the light increases linearly with current. In practical cases, the spontaneous emission (finite β) increases the output at low currents and, at the higher current levels, other optical modes of the injection laser may become excited, and such additional modes are shown up, not only on the emission patterns of the fields, but also in the light/current characteristics by slight kinks or deviations from linearity.

6.4.2 *Small signal deviations*

With the laser biased with a steady current $I_o > I_t$, the photon density is P_o and the electron density is n_t. A small r.f. current I_1 is superimposed on I_o leading to changes P_1 and n_1 in P and n. Neglecting products $P_1 n_1$ and using $n_t - n_o = 1/G\tau_p$

$$dP_1/dt = GP_o n_1 \tag{6.4.5}$$

$$dn_1/dt = -(n_1/\tau_r) - GP_o n_1 - (P_1/\tau_p) + D_1 \tag{6.4.6}$$

Note that whenever n_1 is zero, P_1 is a maximum or minimum, so changes in P vary in quadrature with changes in n (problem 6.6). Eliminating n_1:

$$(d^2P_1/dt^2) + 2a(dP/dt) + \omega_r^2 P = KD_1 \tag{6.4.7}$$

where

$$2a = (1/\tau_r) + GP_o;\ K = GP_o$$

and

$$\omega_r^2 = GP_o/\tau_p = [(I - I_t)/I_t][n_t/(n_t - n_o)][1/\tau_r\tau_p]$$

A small step change in the current drive to the laser creates a response in the light output of

$$\delta L = L_o[1 - \exp(-at)\cos\omega_r t]$$

exhibiting ringing. The frequency response (Fig. 6.9) is important for communications, and the 'photon–electron resonance' provides a limitation in the speed of modulating a laser efficiently. However, with careful design, lasers are able to be modulated with microwave currents up to 10 GHz or so. High levels of spontaneous emission, effects of many optical modes and transverse diffusion of charge carriers can all help to

Fig. 6.8. Light/current characteristic for laser (*a*) From ideal steady state rate equations (rough practical values for a 'communication' laser). (*b*) Allowing for finite spontaneous emission. (*c*) New transverse mode excited at higher currents. (*d*) Ideal spectrum. (*e*) With additional longitudinal modes. (Note threshold currents for diode lasers are forever decreasing as technology improves. A few milliamperes is possible.)

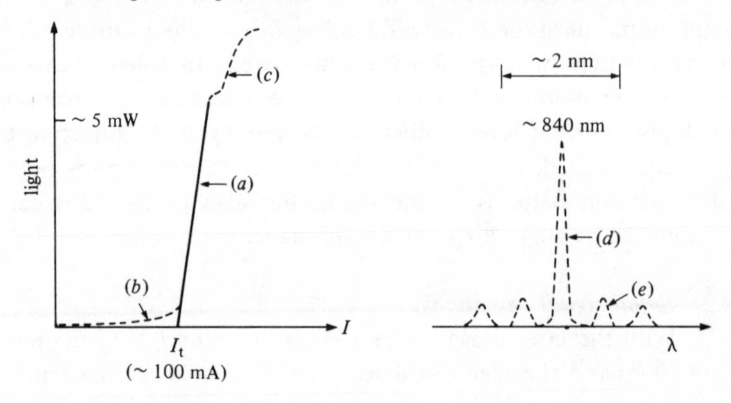

alter the resonance so that the detailed shape of the frequency response varies from device to device.[63]

6.4.3 *Large signal gain switching*

The large signal effect of this photon–electron resonance is best computed. Fig. 6.10 shows one example with selected parameters where the current has been suddenly stepped from $0.9I_t$ to $1.75I_t$. The ringing is very marked, and indeed by simply driving the laser with a pulse of current about 200 ps duration, shorter single optical pulses can be generated, with the light pulses essentially being the initial 'ring'. The shorter the diode, then the shorter is the photon lifetime and the shorter the time scale for the light generation. With such current drives, optical pulses of 15 ps duration have been produced from 50 μm long lasers. With optical pumping, pulses 1 to 2 ps duration have been produced with material a micron or so thick. Effects of mode locking within the laser cavity of the diode may also be expected to play a role (see section 7.4). Special lasers with a high modulation capability must also be expected to give shorter pulses.

This method of pulse production is called gain switching. The gain is switched by the driving pulse from absorption to gain and the stimulated emission depletes the electron population switching the gain back to absorption. It may be difficult to use this technique in a high quality communication system because of the change of electron density which modulates the refractive index of the laser guide by a small amount and so alters the precise frequency at which the device emits light. This optical

Fig. 6.9. Schematic modulation frequency response for laser and LED. The resonance angular frequency ($\sim 10^{10}$ rad/s) tends to limit response which falls at 40 dB/decade after this resonance. In the LED, the recombination angular frequency ($1/\tau_r \sim 10^9$) tends to limit response which falls off then as 20 dB/decade. The package for the device and other details can change the response from this ideal.

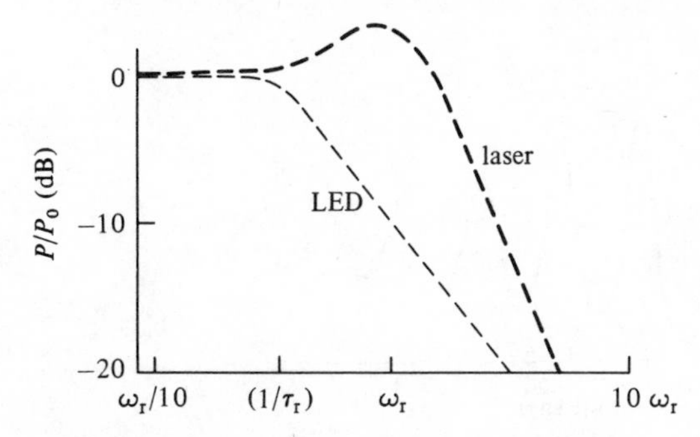

frequency 'chirp' is undesirable in communications and needs to be eliminated.

6.5 Rate equations for the light emitting diode

6.5.1 *Steady state conditions*

Idealised rate equations for an LED similar to those for a diode laser are outlined here, again building on (6.2.13) and (6.2.16) but now assuming that all the spontaneous emission couples into the broad range of radiative emission of density P, which gives an average optical gain $PG(n-n_o)$ over the spectrum. Thus, unlike the laser, the photon density P no longer refers to a single wavelength but to a range of photon energies typically about $2kT$ wide. The rate equations then read

$$dP/dt = GP(n-n_o)+(n/\tau_{rr})-(P/\tau_p) \quad (6.5.1)$$

$$dn/dt = -GP(n-n_o)-(n/\tau_{rr})-(n/\tau_{nr})+D \quad (6.5.2)$$

where D is the drive $(I/q\Phi)$ and we have allowed for both radiative and non-radiative recombination (τ_{rr}, τ_{nr}) to demonstrate the effects on the internal quantum efficiency η_i.

Setting $dP/dt = dn/dt = 0$ and rearranging (6.5.1) and (6.5.2):

$$(I/q) = L\{1+(\tau_{rr}/\tau_{rn})[\Phi n_t/(L\tau_{rr}+\Phi n_t-\Phi n_o)]\} \quad (6.5.3)$$

where $L=\Phi P/\tau_p$ photons/s is the total theoretically available light

Fig. 6.10. Photon–electron resonance. Elementary rate equation estimation of photon electron resonance for laser with 2 ns recombination time and 2 ps photon lifetime with transparency density $0.8n_t$. Step drive: 0.95 to $1.75I_t$. Practical responses generally show higher damping, which may be caused by gain saturation, higher spontaneous coupling and other factors.

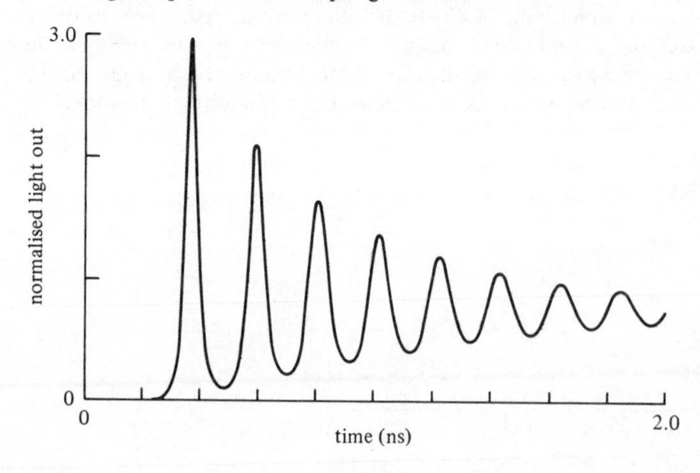

output and $n_t = n_o + (1/G\tau_p)$ gives the notional value of electron density at which the stimulated emission could overcome the large losses and set the device lasing. This density n_t is now much higher than for a properly constructed laser and lasing does not occur. The light output L is approximately linear with drive current, at least at low levels. The internal quantum efficiency η_i is given by

$$L = \eta_i(I/q) \tag{6.5.4}$$

where

$$\eta_i = \tau'_{rn}/(\tau'_{rn} + \tau_{rr})$$

with

$$\tau'_{rn} = \tau_{rn}[(n_t - n_o)/n_t]$$

Non-radiative recombination then, as expected, reduces the internal quantum efficiency according to (6.5.4). There are internal losses by absorption of photons in non-generating regions which also reduce the internal efficiency. Of course it is not possible to collect all the light that in principle is generated, especially as it comes out in many directions. This leads to a coupling efficiency around 50%, with care, and an overall efficiency around 10%.

The LED offers several important differences compared with a laser. First the light output is linear with the drive having no threshold current (Fig. 6.11), and linearity is usually an advantage in practical systems. The spectral output covers a range of wavelengths and so couples to several modes in a fibre. Such coupling has advantages and disadvantages. The disadvantages are that the energy in the different wavelengths and modes travel at slightly different velocities so spreading any pulse in time and limiting the speed of modulation over a long length of optical fibre.

Fig. 6.11. Current/light characteristics for LED. (*a*) I/L. (*b*) Spectrum. (Values give rough practical guides for GaAs LEDs.)

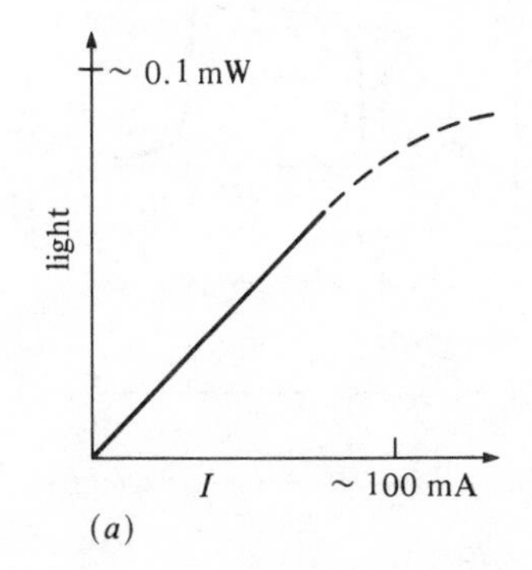

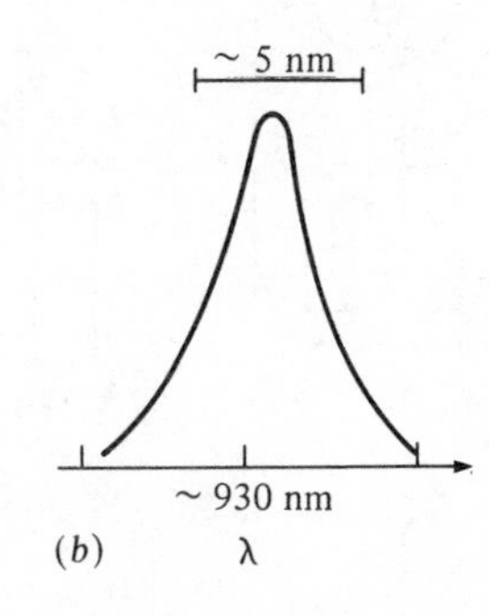

On the other hand, the spread in wavelengths smooths out the strong interference patterns which can be found for 'single' wavelength lasers when the outputs couple to two or more modes (modal noise[66]).

6.5.2 *Dynamic performance*

Assuming that the electron density remains well below the notional transparency value n_o, the equations linearise to give

$$(\mathrm{d}P/\mathrm{d}t)+(P/\tau_p')=n/\tau_{rr} \tag{6.5.5}$$

$$(\mathrm{d}n/\mathrm{d}t)+(n/\tau_r)=D-Gn_oP \tag{6.5.6}$$

where

$$1/\tau_p'=Gn_o+(1/\tau_p)$$

and

$$1/\tau_r=(1/\tau_{rr})+(1/\tau_{rn})$$

The recombination time is so much longer than the photon lifetime that in general the latter can be neglected, giving an approximate dynamic response from

$$\tau_r(\mathrm{d}L/\mathrm{d}t)+L=\eta(I/q) \tag{6.5.7}$$

For a current drive modulated as $\exp j\omega t$ the modulation response is given by Fig. 6.10, with no resonance frequency.

PROBLEMS 6

6.1 A nearly ideal photodiode feeds a 50 Ω resistive load and has 0.5 pF capacitance (Fig. 6P(a)). The transit time of the charge carriers across the depletion zone has an average of 25 ps from the generation position. Estimate the frequency response of V

Fig. 6.P.

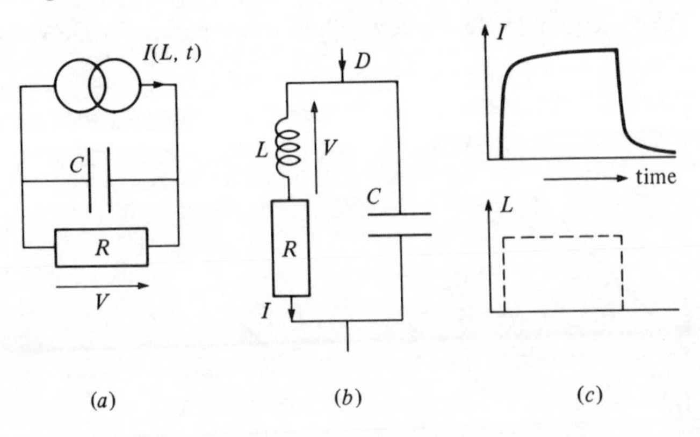

to variations of the light input. Estimate the frequency for the -3 dB point in the response. Bias problems, stability etc. can be ignored for the discussion here. Hint: neglect n/τ_r in (6.3.3).

6.2 If the voltage across a diode exceeds breakdown by 30%, use the rate equation (6.3.8) (take $m=7$) to estimate the rate of growth of the charge carriers by impact ionisation. Take the avalanche zone as effectively 0.1 μm thick with a scattering limited velocity of 10^5 m/s.

6.3 For Fig. 6P(b), show that $C(\mathrm{d}V/\mathrm{d}t)+(CR/L)V+I=D$, with $L(\mathrm{d}I/\mathrm{d}t)=V$. Compare these equations with the linearised laser equations (6.4.5) and (6.4.6), showing this circuit is an equivalent circuit for the laser with a light output $P\Phi/\tau_p=I/e$, where I is the current through a resistance R and $CV=en_1\Phi$, the stored electronic charge in the laser. Evaluate R and L.

6.4 Given an additional internal loss $-P/\tau_{pi}$ in (6.4.1), show that the single facet laser efficiency is reduced to

$$[(I-I_t)/2I][\tau_{pi}/(\tau_{pi}+\tau_p)]$$

For a laser driven at twice threshold, the single facet efficiency is found to be 12.5% ; find (τ_{pi}/τ_p).

6.5 In a particular p-i-n photodiode, the contacts are found to absorb 25% of the light so that there are two components n and n_c for the photogenerated electrons. The contact electrons (n_c) diffuse into the i-region at a rate n_c/τ_{rc}, where τ_{rc} is a time on the scale of 0.5 ns. The recombination time is 10 ns for carriers, n, in the i-region. The effective electron transit time τ across the i-region is 50 ps. Explain the terms in the two rate equations

$$\mathrm{d}n/\mathrm{d}t=(3P/4)Gn_o-(n/\tau_r)-(n/\tau)+(n_c/\tau_{rc})$$

$$\mathrm{d}n_c/\mathrm{d}t=(P/4)Gn_o-(n_c/\tau_{rc})$$

Explain the response of Fig. 6P(c) to an ideal square input of light to this diode.

6.6 For the result of Fig. 6.10, sketch qualitatively the electron density as a function of time, using (6.4.5).

7

Advanced topics in rate equations

7.1 Introduction

Further exciting developments in rate equations are possible. Some of these more advanced uses and techniques are touched on in this chapter, singling out laser devices which will find applications in communications. The statistical information that can be provided by rate equations has been one theme in this book and is developed further to demonstrate how the output of a single mode injection laser changes from a chaotic distribution to a Poisson distribution as the drive current into the laser is increased.[54]

The injection laser normally has several modes.[63,67–69] It is useful then to show how rate equations can handle such multimode problems. In particular this section emphasises the importance of spontaneous emission in determining mode amplitudes.

Rate equations, as interpreted here, have been concerned with rates of change of energy, momentum, quanta, charge and so on. These equations have all removed any information of the phase of quantum or electromagnetic waves. In phenomena where phase is important more detailed discussions using full quantum or electromagnetic theories are usually required. To demonstrate the importance of phase and also to demonstrate how the rate equation approach can sometimes be modified to include phase, the 'mode locked' laser is discussed briefly. This topic follows on naturally from the multimode rate equations because in a mode locked laser[70–73] there are many optical modes at equally spaced frequency intervals but with their amplitudes locked to zero phase at one time. The resultant output from such mode locked lasers can be a train of exceptionally short pulses, down to subpicosecond durations with nanosecond repetition rates.

The chapter is concluded with a fundamental discussion developing photon rate equations from Maxwell's equations.[74–76] This classical view gives insight into the development of rate equations and the quantum/classical link for photons.[77] It is of practical help in demonstrating links with material properties and the general principles of energy storage in stabilising oscillators,[76] a result that is valid over the whole spectrum from audio oscillators to lasers.

7.2 Photon statistics for injection lasers

7.2.1 *Recasting the laser equations*

The analysis for the photon distribution in a single mode injection laser starts by recasting the idealised model of a semiconductor given in sections 5.5 and 6.2, where, in a unit volume of the interaction region, there are effectively N_c states containing n_c electrons in the conduction band, and for the valence band there are effectively N_v states containing n_v electrons (i.e. $N_v - n_v$ holes).

Recasting the model of Fig. 6.4, define occupation probabilities from

$$n_c/N_c = r_c; \quad n_v/N_v = r_v \tag{7.2.1}$$

For the laser in equilibrium, $n_c = n_{ce}$, $n_v = n_{ve}$ or $r_c = r_{ce}$ and $r_v = r_{ve}$.

The photon rate equation, including spontaneous emission, is given from (5.3.7) suitably modified to

$$dP/dt = (P+1)g_o r_c(1-r_v) - Pg_o r_v(1-r_c) - (P/\tau_p) \tag{7.2.2}$$

where P are the total number of photons in the uniform interaction of volume Φ and the gain is $g_o = GN_cN_v$. In (7.2.2), the first term gives the stimulated and spontaneous emission ($\propto P+1$), the second term gives the absorption ($\propto P$), while the final term gives the rate of loss (lifetime τ_p) of photons to the external system (i.e. the potentially useful photons).

Because of subpicosecond dielectric relaxation time for the plasma of recombining holes and electrons, electrical charge neutrality, which has previously been assumed in these lasers, is also to be assumed here. For an ionised acceptor density of N_A in the recombination region, this neutrality must give

$$p = N_v(1-r_v) = n_c + N_A = N_c r_c + N_A$$

or

$$1 - r_v = Rr_c + S \tag{7.2.3}$$

where

$$R = N_c/N_v: \quad S = N_A/N_v < 1$$

The equilibrium photon numbers are then given from

$$P_e = r_{ce}[S + Rr_{ce}]/[1 - S - (1+R)r_{ce} + 1/g_o\tau_p] \qquad (7.2.4)$$

As the photons increase, the electron density reaches a limiting value

$$r_{ce} \to r_{ct} = [1 - S + 1/g_o\tau_p]/[1+R] \qquad (7.2.5)$$

In the analysis where spontaneous emission was neglected, no emission occurred until the electron density reached the 'threshold value' n_{ct}. Here $n_c < n_{ct}$ for the steady state. The gain g_o and loss rate $1/\tau_p$ will not be too different in magnitude so that it is helpful to define

$$G_o = g_o\tau_p \qquad (7.2.6)$$

where G_o is of order unity.

The electron rate equation may be written as

$$\Phi N_c(dr_c/dt) = -(P+1)g_o r_c(1-r_v) + Pg_o r_v(1-r_c) - (N_c\Phi r_c/\tau_r) + D \qquad (7.2.7)$$

where D $(=I/q)$ is the drive (electrons/s) into the interaction volume Φ, and τ_r is the spontaneous recombination time for those electron-hole recombinations which do not contribute to the lasing photons P. The recombination time τ_r for an injection laser is the order of nanoseconds, while the photon lifetime τ_p is around a picosecond, so, writing $\tau_r = Q\tau_p$, $Q \sim 1000$.

Define quantities

$$P_c = \Phi N_c/Q; \qquad d = D\tau_p \qquad (7.2.8)$$

so that (7.2.2), (7.2.7) and (7.2.8) give a further equilibrium condition

$$P_e = d - P_c r_{ce} \qquad (7.2.9)$$

It may be shown that when the drive d is $d_t = P_c r_{ct}$ then P_e is the order of $(P_c)^{1/2}$, while, if d is kd_t with $k \gg 1$, then $P_e \sim (k-1)P_c r_{ct}$. Consequently, with spontaneous emission included, $d = d_t$ is effectively the drive threshold for lasing, just as it is with spontaneous emission neglected.

7.2.2 *Rates of change of probability*

As for section 5.5, define E_P as the probability of finding a state of the photon field with P photons. Then, following Loudon,[54] let R_{cP} be the probability of finding the conduction band having an occupation probability $r_c = R_c$ simultaneously with there being P photons present. Similarly let R_{vP} be the probability of finding the value R_v for the occupation probability r_v of the valence band simultaneously with P

photons in the field. To satisfy the requirements of probability:

$$\sum_P \{E_P\} = 1; \quad \sum_P \{PE_P\} = P_e \tag{7.2.10}$$

$$N_c \sum_P \{R_{cP}\} = n_{ce}; \quad N_v \sum_P \{R_{vP}\} = n_{ve} \tag{7.2.11}$$

where n_{ce}, n_{ve} and P_e are the equilibrium, steady state values.

Section 5.5 shows how to find the increase in the probability E_P by noting that if the field starts with an occupation value of P, then emission, absorption and loss from the cavity change the photons to $P+1$ or $P-1$ and so must reduce the probability E_P. Starting from $P+1$ photons, absorption and loss of one photon brings the state back to the P state and so increases the probability E_P. Equally, starting from the $P-1$ state, an emission brings the system back into the P state, again increasing the probability E_P. Following this through term by term leads to

$$\begin{aligned}\tau_p \, dE_P/dt = &-(P+1)G_o[R_{cP} - (R_cR_v)_P] - PG_o[R_{vP} - (R_cR_v)_P] \\ &- PE_P + (P+1)G_o[R_{v\langle P+1\rangle} - (R_cR_v)_{\langle P+1\rangle}] \\ &+ (P+1)E_{\langle P+1\rangle} \\ &+ PG_o[R_{c\langle P-1\rangle} - (R_cR_v)_{\langle P-1\rangle}]\end{aligned} \tag{7.2.12}$$

where $(R_cR_v)_P$ is the probability of finding both values $r_c = R_c$, $r_v = R_v$ simultaneously with P photons present.

The change of electron occupation probability is decreased by an emission from the P photon state with a value R_{cP} but is increased by an absorption starting from the $P+1$ photon state, with a value $R_{v\langle P+1\rangle}$, turning into a P state. The drive is assumed to be constant so that the probability of a value D for the drive simultaneously with P photons is just E_PD. Hence:

$$\begin{aligned}\Phi N_c \tau_p \, dR_{cP}/dt = &-(P+1)G_o[R_{cP} - (R_cR_v)_P)] \\ &+ (P+1)G_o[R_{v\langle P+1\rangle} - (R_cR_v)_{\langle P+1\rangle}] \\ &- P_cR_{cP} + E_Pd\end{aligned} \tag{7.2.13}$$

The equivalent equations for the occupation of the valence band would be similarly given from the stimulated emission and absorption as:

$$\begin{aligned}\Phi N_v \tau_p \, dR_{v\langle P+1\rangle}/dt = &(P+1)G_o[R_{cP} - (R_cR_v)_P] \\ &-(P+1)G_o[R_{v\langle P+1\rangle} - (R_cR_v)_{\langle P+1\rangle}] \\ &+[P_cR_{v\langle P+1\rangle}/R] \\ &+(\text{drive and space charge terms})\end{aligned} \tag{7.2.14}$$

The equation has deliberately been written in terms of $dR_{v\langle P+1\rangle}/dt$ to show the similarity of the result with $-dR_{cP}/dt$. Space charge is the term used for distributed charge throughout any device which leads to additional electric fields. It is these space charge fields which move the charge so as to restore overall electrical charge neutrality on a time scale of the dielectric relaxation time. Recombination of holes and electrons in pairs also helps to preserve neutrality so that a small adjustment in drive, through the undetermined space charge fields, can give:

$$S_1 E_{\langle P+1 \rangle} - R_{v\langle P+1 \rangle} = RR_{cP} + S_2 E_P \tag{7.2.15}$$

where S_1 and S_2 are chosen for self consistency. If

$$S_1 = R_{v0}/E_0 \quad \text{and} \quad S_2 = S + S_1 - 1 \tag{7.2.16}$$

then summing (7.2.15) over all values of P yields precisely (7.2.3). (7.2.15) and (7.2.16) are therefore equivalent to the approximation for electrical neutrality.

7.2.3 *Equilibrium photon distributions*

The dynamic photon distributions lie outside the scope of this text, but the steady state photon distributions can be found with a little further work.

(7.2.12) and (7.2.13) taken together show that, with $\mathrm{d}/\mathrm{d}t = 0$,

$$P_c R_{cP} - E_P d + (P+1) E_{\langle P+1 \rangle} = P_c R_{c\langle P-1 \rangle} - E_{\langle P-1 \rangle} d + P E_P \tag{7.2.17}$$

Any relationship of the form $F(P) = F(P-1)$ implies that the function F is a constant, independent of P. (7.2.17) is in this form, and at the value of $P = 0$ the recombination would match the drive current so that for all P

$$P_c R_{cP} - E_P d + (P+1) E_{\langle P+1 \rangle} = 0 \tag{7.2.18}$$

with $P_c R_{c0} = E_0 d$. It may be checked, by summing over all photon states P, that (7.2.18) correctly gives (7.2.9).

To evaluate a second relationship between E_P and R_{cP}, (7.2.12) is used, with $\mathrm{d}/\mathrm{d}t = 0$, written approximately as

$$\begin{aligned} 0 = &-(P+1) G_o R_{cP}(1 - r_{ve}) + (P+1) G_o R_{v\langle P+1 \rangle}(1 - r_{ce}) \\ &- P_c R_{cP} + E_P d \end{aligned} \tag{7.2.19}$$

With (7.2.15) and (7.2.19), and some rearrangement

$$E_P(d + P_{c2}) = E_{\langle P+1 \rangle}(P + 1 + P_{c1}) \tag{7.2.20}$$

where

$$P_{c1} = P_c[1 + G_o(1 - r_{ce}) S_1] / G_o[1 + R - r_{ve} - R r_{ce}]$$

and

$$P_{c2} = P_c(1 - r_{ce}) S_2 / [1 + R - r_{ve} - R r_{ce}]$$

Hence

$$E_P = E_0[d+P_{c2}]^P[\Gamma(P_{c1})/\Gamma(P+P_{c1})] \tag{7.2.21}$$

where Γ gives the gamma function: $\Gamma(N)=(N+1)!$ for N integer, and E_0 is determined from the requirements of (7.2.10).

Taking $P_{c1} \sim P_c \sim d_t$ and $P_{c2} \sim 0$, then if the drive d is large enough so that $P \gg P_c$, the photon distribution approximates to a Poisson stream with a mean $P = d - P_c$. If the drive d is small enough then the output approximates to that of a single mode chaotic distribution (Fig. 5.8) with a mean $1/(1-d)$ ($d \ll 1$, $P_{c2}=0$).

7.2.4 *Practical consequences*

The main significance of this section is to emphasise the important tie between rate equations and temporal statistics, just as the tie between rate equations and energy statistics was revealed earlier.

The technological implications for a good design of communication injection laser will be considered by future designers of advanced systems using coherent optics where a good Poisson stream of photons at a single frequency will be used. The calculation here indicates, for a large mean photon number (n_{mean}) and 'threshold' photon count P_{c1}, that one obtains a nearly gaussian distribution with a variance of $(n_{\text{mean}}+P_{c1})^{1/2}$ rather than the correct gaussian approximation to the Poisson distribution where the variance is $(n_{\text{mean}})^{1/2}$. Operating a laser at 10% above threshold would give a variance $(\frac{11}{2})^{1/2}$ times a laser which had the same photon count but was operated at twice threshold. It may then be that the further development of single mode low threshold lasers which can operate several times above the threshold level will be advantageous.

7.3 **Multimode rate equations for injection lasers**

7.3.1 *Laser modes*

The discussion for lasers has been mainly about single mode devices, and little has been said about how the laser selects the mode in which it operates. Practical lasers frequently exhibit several modes to some degree. Rate equation analysis shows some of the important factors affecting this problem of multimodes and it appears as an appropriate topic in more advanced techniques.

An elementary account of electromagnetic modes for plane waves in an ideal Fabry–Perot resonator (length L) is given in appendix A, where standing waves $\exp(j\omega t)\sin kz$ are permitted with $kL=2m\pi$ and m is integer. For plane waves $k=\omega/c_m$, where c_m $(\sim c/\varepsilon^{1/2})$ is the phase velocity within the medium forming the resonator. In a laser there is typically a central mode at a frequency ν_0 with other modes, frequency ν_m, on either side:

$$\nu_0 = m_0 c_m / L; \qquad \nu_m = \nu_0 \pm m\nu_r \tag{7.3.1}$$

where, if c_m is independent of frequency, $\nu_r = c_m/L$ independent of the integer *m*. The central mode is determined not just by the resonator but also where the optical gain of the medium is strongest as will be shown here.

In a real laser, the details are modified by transverse variations of the field pattern which depart from true plane wave. The details of waveguiding with transverse modes must be left for further reading,[63] but the principle result of (7.3.1) remains as a useful guide for the work here.

7.3.2 *Multimode: steady state conditions*

The model of section 6.4 for the injection laser with a single mode may be extended directly. First define P_m as the photon density for the *m*th mode. All the electrons (density n) can contribute to the stimulated emission for all these modes because, as explained before, it is assumed that electrons can 'thermalise' or redistribute their energies *within* the conduction band in picosecond time scales. Thus if one electron energy becomes more depleted than neighbouring energies, the electrons change state rapidly so as to refill this depleted mode. For processes taking longer time scales than this thermalisation it is not necessary to have separate rate equations determining the movement of electrons within the conduction band. This approximation has been found to be of practical use. Similar remarks apply for the valence band. In this subsection, the steady state conditions are discussed to find the number of modes excited. The formalism set up for the single mode photon equations then holds and one may write for each mode

$$\mathrm{d}P_m/\mathrm{d}t = G_m P_m (n - n_{om}) - (P_m/\tau_{pm}) + (\beta_m n/\tau_r) \tag{7.3.2}$$

where in principle one may allow for slightly different gains G_m, transparency densities n_{om}, photon lifetimes τ_{pm}, and spontaneous emission coupling factors β_m for each mode.

The stimulated recombination of the electrons has all the photons working together so that the rate equation for the electron density becomes

$$\mathrm{d}n/\mathrm{d}t = -\left\{\sum_m [G_m P_m (n - n_{om})]\right\} - (n/\tau_r) + (I/e\Phi) \tag{7.3.3}$$

The mode where the lasing material provides the highest gain G_0 is taken as $m = 0$. The modes on either side have a lower gain, which is assumed to be given by

$$G_m = G_0/[1+\gamma m^2] \tag{7.3.4}$$

with an appropriate value of γ.

In the steady state, with $\mathrm{d}/\mathrm{d}t = 0$

$$P_m = \beta_m n/\{\tau_r[G_m(n-n_{om})-(1/\tau_{pm})]\} \tag{7.3.5}$$

Now define n_t, γ' and δn from

$$\begin{aligned} n_t &= n_{oo} + 1/G_0\tau_{p0}; \quad \delta n = (n-n_t)/n_t \\ n_t\gamma' m^2 &= n_{om} - n_{oo} + (1/G_m\tau_{pm}) - (1/G_0\tau_{p0}) \end{aligned} \tag{7.3.6}$$

Changes in the photon lifetime with optical wavelength can be engineered by coating the facets of the laser with frequency selective dielectric films, by having additional optical cavities, or by other appropriate changes. All the variations in the gain, transparency density and photon lifetime have been lumped together in a single parameter giving an overall curvature from γ'. If only the material properties varied the gain G_m then

$$n_t\gamma' = (n_t - n_{oo})\gamma \tag{7.3.7}$$

In general the variations of the gain are not strongly wavelength dependent ($\gamma \ll 1$) and n is close to n_t around threshold so that one can write (7.3.5) approximately as

$$P_m = \beta_m/[\tau_r G_0(\delta n + \gamma' m^2)] \tag{7.3.8}$$

An immediate result can be seen (Fig. 7.1) that with spontaneous emission included, a high stimulated light output is obtained by having δn small. The overall gain curvature of the combined material and cavity then affects the excitation of other modes compared to that with the highest gain.

Now because δn is small, $[1+(\gamma/\delta n)m^2]$ changes more rapidly with m than τ_m/τ_0. Further, $\sum_m \{\beta_m\} \ll 1$ so that (7.3.2) and (7.3.3) may be combined to give

$$\sum_m \{P_m/\tau_{p0}\} \equiv (I/e\Phi) - (n/\tau_r) \tag{7.3.9}$$

The summation in (7.3.9) using (7.3.8) is achieved by noting

$$\sum_m \{1/(am^2+b)\} = \pi(1/ac)\coth(\pi c) \tag{7.3.10}$$

where the sum is from $-\infty$ to $+\infty$ and

$$c^2 = b/a$$

Hence assuming a uniform value for the spontaneous emission and small δn

$$\pi(1/\gamma'\delta n)^{1/2}\coth\{\pi(\delta n/\gamma')^{1/2}\} = [\delta n + (I-I_t)/I_t][n_t/\beta(n_t - n_o)] \tag{7.3.11}$$

The number B of excited modes can be taken approximately from where P_m falls to $P_0/2$ in (7.3.8), so taking $B = m_1 - m_2$, where $m_i^2\gamma'/\delta n = 1$

$$B = 2(\delta n/\gamma')^{1/2} \tag{7.3.12}$$

Even for $B = 1$, the coth function in (7.3.11) is approximately unity, so that with $I > I_t$ and (7.3.7), the number of modes excited is given approximately from

$$B = \{2\pi\beta/\gamma[(I - I_t)/I_t]\} \tag{7.3.13}$$

This result shows that good single mode operation requires the spontaneous emission coupling to be low, combined with an appropriate fall in the gain (or increase in cavity loss) away from the central mode. As the laser is driven harder so the spectrum should reduce in width. This is well documented in practical work with diode lasers.[67-69] In practice, one sometimes finds that at the higher currents additional transverse modes are excited complicating the issue, but good laser designs attempt to avoid this problem. Injection lasers are being designed to have cavities where the loss varies markedly with frequency through adding external frequency selective elements (see section 7.4) or through including internal structures (such as shown later in Fig. 7.6 - distributed feedback lasers).[63] Similar principles apply to these lasers and excellent single mode operation can be found.

Fig. 7.1. Change of relative excitation of laser modes with electron density $\delta n = n_t - n$. As $n \to n_t$ central mode with highest gain tends to largest output dominating all other modes. Mode amplitude is proportional to coupled spontaneous emission and $1/\delta n$.

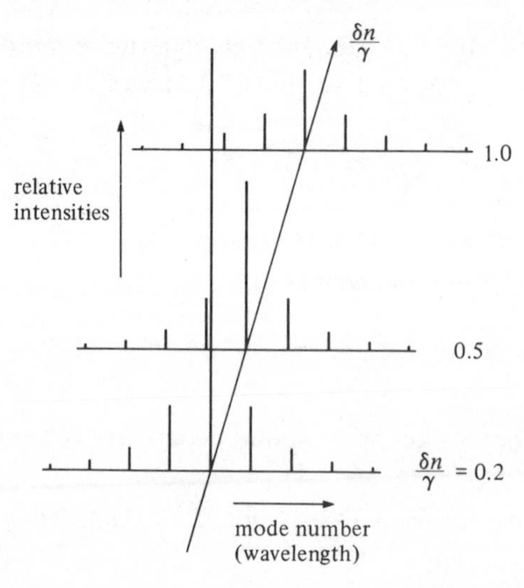

7.3.3 *Multimode dynamics*

The approximate dynamics of multimode lasers are of interest because it is found experimentally that for short enough pulses a laser tends to have more modes excited than in the steady state. One needs to know the time scale on which the laser becomes single mode.

It is helpful to recognise that nothing much happens unless $n \sim n_t$, the limiting steady state electron density. Writing

$$n = n_t + B(t)(n_t - n_o) \tag{7.3.14}$$

and taking the photon lifetimes and transparency densities to be the same for all the neighbouring modes, one can write

$$\mathrm{d}P_m/\mathrm{d}t = (P_m/\tau_p)(B - \gamma m^2) \tag{7.3.15}$$

A solution to this equation is given by

$$P_m = A_m F(t) \exp(-\gamma m^2 t/\tau_p) \tag{7.3.16}$$

with $F = \exp\{-\int_0^t [B(t')/\tau_p'] \, \mathrm{d}t'\}$ independent of the mode and consequently not affecting the relative modal intensities. The values of A_m are determined by the initial modal intensities dependent upon the initial spontaneous spectrum of emission below threshold. Assuming a gaussian spread in that spectrum at $t = 0$

$$P_m = A_o F(t) \exp[-(m^2/2M^2)(1 + 2M^2\gamma t/\tau_p)] \tag{7.3.17}$$

The standard deviation of the number of modes excited then narrows with time according to

$$M(t)/M = 1/(1 + 2M^2\gamma t/\tau_p)^{1/2} \tag{7.3.18}$$

narrowing to a single mode (Fig. 7.2) when $t \sim \tau_p/2\gamma$.

This result has practical consequences for pulsed lasers. If the lasing material is suddenly switched into a strong positive gain, then many modes will be excited and start to grow because $n > n_t$ for all the modes.

Fig. 7.2. Schematic change of modal excitation with time for pulse drive narrowing to ideal single mode.

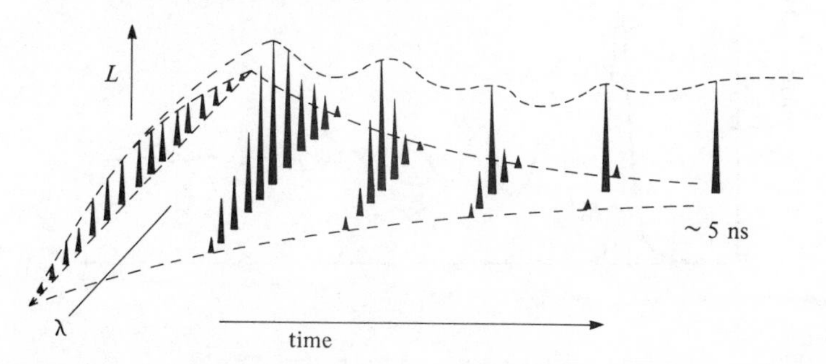

However, if the effective gain changes markedly with mode, the multimode operation decays rapidly with time into a single mode. Again external cavities or distributed feedback which increase the losses for unwanted modes will help maintain single mode operation even for subnanosecond pulses.

The total light output still exhibits the ringing and resonances associated with the single mode output. Forming the sum of the modal powers, $P=\sum_m \{P_m\}$. Then, with gains varying slowly from mode to mode, approximate rate equations are

$$dP/dt = G_{\text{mean}}P(n-n_o)-(P/\tau_p)+\sum\{\beta n/\tau_r\} \qquad (7.3.19)$$

$$dn/dt = -G_{\text{mean}}P(n-n_o)-(n/\tau_r)+D \qquad (7.3.20)$$

The major difference between the approximate result for many modes and the single mode is the increase in the spontaneous emission through summation of spontaneous power. The dynamics follow as for the single mode event, and Fig. 7.3 shows the result of increasing the spontaneous coupling factor by 5 from Fig. 6.11. This demonstrates the feature that high spontaneous emission levels dampen the photon–electron resonance.

Further detailed analysis with different thresholds, spontaneous emission coupling, transparency densities, along with transverse distributions of the field and so on all affect the precise answers for the amount of damping and resonance. But this detail is left for the specialist.[63]

Fig. 7.3. Step drive to laser with additional spontaneous emission appropriate for many modes. (Same parameters as for Fig. 6.4 except five times higher total spontaneous emission coupled into combined lasing modes has been assumed.)

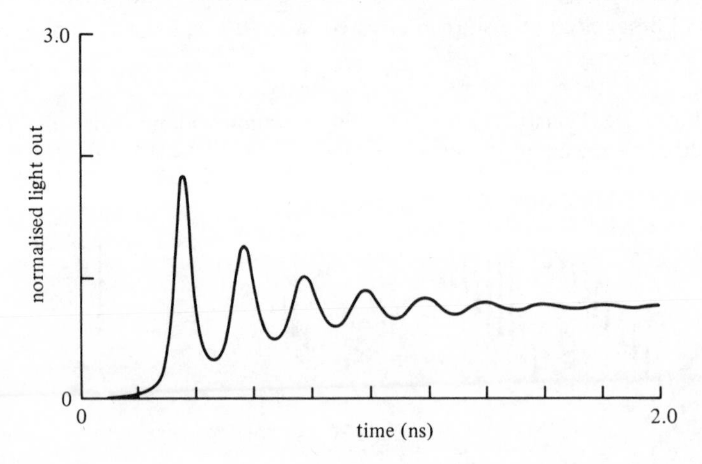

7.4 Rate equations with phase included

7.4.1 *Introduction to mode locking in lasers*

An approximate solution of interactions between electrons and photons using rate equations is often simpler than a solution using full quantum and electromagnetic theories. A penalty for the simplicity is the lack of phase information about any quantum or electromagnetic wave. This is a significant limitation on the use of rate equations. To demonstrate the importance of phase, and also to show that sometimes it is possible to adopt a rate equation formalism which includes phase, the problem of a multi-mode laser is examined approximately and the phenomena of mode locking analysed. This work follows on from section 7.3 and outlines some of the physics of a method which had produced subpicosecond optical pulses.[73]

Mode locking can be explained by applying Fourier analysis to the structure of laser modes. Starting with the model for the Fabry–Perot resonator (see (7.3.1)) it was seen that there were modes separated by angular frequencies $\omega_r = 2\pi c_m / L$. Given a medium of sufficiently broad optical gain, it has also been shown how spontaneous emission may excite many modes. In general, the amplitude of the optical output is of the form

$$A(t) = \sum_m \{A_m \exp[j(m\omega_r + \omega_o)t]\} \tag{7.4.1}$$

where $\omega_o = m_o\omega_r$ gives the central frequency and the summation is over all the mode amplitudes A_m. Note that $\omega_r = 2\pi/T_r$, where T_r is the round-trip time for light in the optical cavity.

Consider light of amplitude $A(t) \exp j\omega_o t$, where $A(t)$ is a pulse which modulates a central optical frequency $\omega_o/2\pi$. The Fourier transform m[78] of $A(t)$ is given by

$$F(\omega) = \int_{-\infty}^{+\infty} A(t) \exp(-j\omega t)\, \mathrm{d}t \tag{7.4.2}$$

If this pulse $A(t)$ is periodic, period T_r, then the output is the same as for (7.4.1), and, from (7.4.2), $A_m = F(\omega_m)$. For a gaussian pulse:

$$\begin{aligned} A(t) &= B \exp\{-\pi(M_0 t/T_r)^2\}; \\ A_m &= (B/M_o) \exp\{-\pi(m/M_o)^2\} \end{aligned} \tag{7.4.3}$$

The phases of all the mode amplitudes are zero at the reference times $t = 0$, T_r, $2T_r$ etc., so that the modes are said to be phase locked together. The larger the value of M_o, then the more frequencies are significant and the shorter the optical pulse. A standard result from Fourier analysis follows. The spread in frequencies δf is inversely related to the spread in time δt:

$$\delta f\,\delta t \sim \Delta \tag{7.4.4}$$

where $\Delta \approx 2\ln 2/\pi$ for the full width half maximum (FWHM) power/energy values (i.e. the square of the amplitudes in (7.4.3)) of the temporal and spectral components.

From (7.4.4), the shortest possible optical pulse is inversely proportional to the optical bandwidth of the system. Fluorescent dyes are one set of materials which with proper excitation show optical gain over a large bandwidth ($\sim 10^{14}$ Hz). Then with specially constructed 'mode locked' systems these lasing materials can produce optical pulses below $\frac{1}{10}$ ps. Semiconductor injection lasers have a potential for producing picosecond pulses. The problems and possible solutions are discussed briefly later.

Figure 7.4(a) shows the key elements of a mode locking system where, at one end of a Fabry–Perot resonator, there is a short device whose gain (G) can be modulated at the resonator round-trip frequency, ω_r, so as to excite many modes with their phases locked to the modulation. The net gain for each mode must be sufficient to overcome all the losses.

One physical realisation is given in Fig. 7.4(b) where a short diode laser has the light, emitted from one facet, collimated into a plane wave

Fig. 7.4. Mode locking systems. (*a*) Schematic system with short region which has gain (modulated at ω_m) coupled to optical resonator. Photon lifetimes τ_R and τ_L give the coupling to right and left, respectively. (*b*) A possible realisation of (*a*) using a diode laser. (*c*) Related scheme but resonator now has gain and the short cavity has a modulated loss.

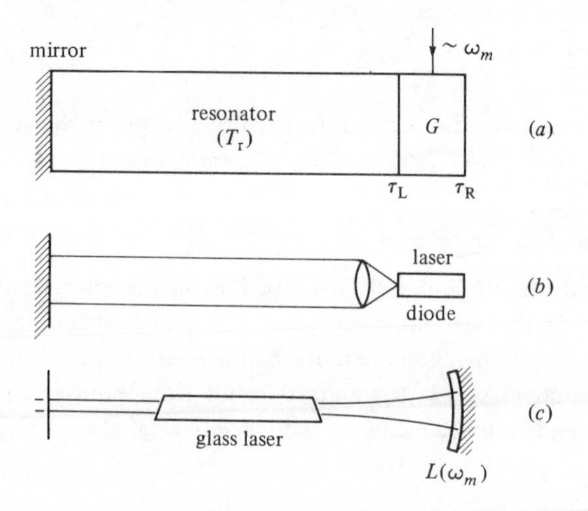

and subsequently reflected back by a mirror into the laser. The mirror forms an external resonator which is coupled to the laser facets. Spherical external mirrors have also been used, but these have a disadvantage that the laser has to be placed close to the centre of curvature of the mirror, so fixing the length of such an external resonator cavity and thus determining an unalterable repetition rate for mode locked pulses.

As an alternative system (Fig. 7.4(c)) the cavity can contain the lasing medium, while the device at the end modulates the loss within the cavity. A clever method uses a cell containing a dye which absorbs light at low light levels, but at higher optical intensities the higher electron energy levels in the material of dye fill up and the absorption ceases. The cell is said to be bleached. Provided this bleaching happens fast enough and recovers fast enough then changes in the light from the laser self modulate the optical loss of the cell. This method is typically used with high power solid state atomic lasers (e.g. Nd-glass) and has the advantage that the loss is automatically modulated at precisely the correct frequency by any pulses which develop. The disadvantage is that it takes time for the pulses to develop from spontaneous emission and there can be a significant variation from pulse to pulse.

This section, following Haus,[71,72] outlines an analysis which combines some of the simplicity of rate equations with a knowledge of phase.

7.4.2 *Amplitude rate equation with coupled cavity*

Taking the model of Fig. 7.4(a), the optical gain is lumped at one end of the system with an active device which is assumed to be short enough so that there is only one main resonance within its gain bandwidth. This active device is assumed to contain a *total* of P photons with a photon amplitude a defined by

$$a^*a = P \tag{7.4.5}$$

This is a quasiclassical analysis so that $a^*a\hbar\omega$ is just the available electromagnetic energy stored within the active device. The central optical frequency (angular) is ω_o so that a rate equation for the amplitude considers the rate of change of phase and the net rate of growth contributed by the various processes. In the classical approximation all stimulated rates of change are proportional to the amplitude a so that one expects

$$\mathrm{d}a/\mathrm{d}t = [\mathrm{j}\omega_o + (G/2) - (1/2\tau_L) - (1/2\tau_R)]a + kb_R + s \tag{7.4.6}$$

Here τ_L is the photon lifetime for photons escaping to the left, τ_R is the photon lifetime for photons escaping to the right, and any stimulated optical gain is measured by an effective gain G (in diode lasers $G \sim$

$g(n-n_o)$). The 'amplitude lifetimes' will be shown to be twice as long as the photon lifetimes. Any spontaneous emission is coupled in as a noise amplitude s, but $s \sim 0$ for most of the discussion. The normalised amplitude of the electromagnetic wave which is to be coupled from the external cavity into the active device is b_R, where $|b_R|^2$ gives the photons/s falling on the interface. The coupling factor k has yet to be determined. Note that $\hbar\omega|b|^2$ gives power flow, in contrast to $\hbar\omega|a|^2$ which gives energy stored. Note b_R represents photons travelling to the right.

The rate of change of phase $j\omega_o$ is determined theoretically by the details of the quantum interaction along with the electromagnetic resonance (see section 7.5). In the style of rate equations, ω_o is assumed to be determined experimentally. Although ω_o is taken to be a constant, this rate of change of phase can be altered, for example as shown later, by the external Fabry-Perot cavity 'pulling' the frequency slightly from ω_o.

With constant gain and no input b_R into the active region

$$a = a_o \exp[j\omega_o + (G/2) - (1/2\tau_L) - (1/2\tau_R)]t$$
$$|a^2| = |a_o^2| \exp[G - (1/\tau_L) - (1/\tau_R)]t \qquad (7.4.7)$$

(7.4.7) reveals the factor of 2 between amplitude and energy (particle) lifetimes. In the steady state, with no growth or decay in the amplitude

$$G - (1/\tau_L) - (1/\tau_R) = 0 \qquad (7.4.8)$$

In the absence of input from b_R, the output from the active device to the left is $|a^2|/\tau_L$ so that

$$|b_L^2| = |a^2|/\tau_L \qquad (7.4.9)$$

showing that the coupling factor between the amplitudes a and b_L is $(1/\tau_L)^{1/2}$. The phase change between the coupled amplitudes is taken to be zero as it would be for an ideal abrupt interface.

The law of reciprocity[74] requires the same coupling factor for the wave, amplitude b_R, feeding back into the active region where the amplitude is a (but see also problem 7.4) giving:

$$k = (1/\tau_L)^{1/2} \qquad (7.4.10)$$

To finalise the coupling equations the total amplitudes on either side of the interface must be equal so that with a coupling coefficient k

$$b_R + b_L = ka \qquad (7.4.11)$$

The phase reference for all the wave amplitudes has been taken at the ideal abrupt interface between the device and external cavity. This result is equivalent to asserting that the total transverse fields E are equal on either side of an interface.

Given a mirror with a reflection coefficient r, and a total delay time T_r for the round-trip time for the photon energy

$$b_R(t+\tfrac{1}{2}T_r)=rb_L(t-\tfrac{1}{2}T_r) \tag{7.4.12}$$

Hence, multiplying (7.4.11) by r and delaying time by T_r

$$rb_R(t-T_r)+b_R(t)=rka(t-T_r) \tag{7.4.13}$$

Now define a time advance operator by

$$A_{Tr}b(t)=b(t+T_r) \tag{7.4.14}$$

Using this operator, (7.4.6) becomes

$$[r+A_{Tr}]\{d/dt-[j\omega_o+(1/2\tau_g)]+(1/2\tau_L)\}a-(r/\tau_L)a=0 \tag{7.4.15}$$

where $1/\tau_g=G-(1/\tau_R)$.

7.4.3 *Solution to amplitude rate equation*

The solution to (7.4.15) needs to start by considering an amplitude $A\exp j\omega t$, so that

$$[r+\exp j\phi][j\omega-j\omega_o-(1/2\tau_g)+(1/2\tau_L)]-[r/\tau_L]=0 \tag{7.4.16}$$

where $\phi=\omega T_r$.

Given a fixed value of τ_g and τ_L, a solution to (7.4.16) is always possible with a complex angular frequency $j\omega=j\omega'+\mu$. The importance of phase is discovered because the frequency (ω' = rate of change of phase) and the growth (μ = rate of change of amplitude) have to adjust to satisfy both the real and imaginary parts of (7.4.16). The initial interest here is to find the correct frequency to give the minimum required gain (minimum power to drive the laser) for a steady state solution ($\mu=0$). On rearranging (7.4.16) one obtains

$$2(j\omega_o-j\omega)+(1/\tau_g)=(1/\tau_L)\{[\exp(j\phi)-r]/[\exp(j\phi)+r]\} \tag{7.4.17}$$

The modulus of the r.h. bracket is given from $\{[1+r^2-2r\cos\phi]/[1+r^2+2r\cos\phi]\}^{1/2}$ showing that the minimum gain (minimum value of $1/\tau$) occurs for $\phi=2n\pi$, with $n=m_o$ an integer for the central mode, where

$$\omega=\omega_o=2m_o\pi/T_r \tag{7.4.18}$$

$$(1/\tau_g)_{min}=(1/\tau_L)(1-r)/(1+r) \tag{7.4.19}$$

The optimum frequency ω_M for the modulation of the gain or loss is determined initially with a fixed gain and is found to be slightly lower than $1/T_r$:

$$\omega_M T_r=2\pi-\delta \tag{7.4.20}$$

The physical reason for this small difference is that the photon interaction in the active region takes a little time so that the period of an ideal periodic process is slightly in excess of T_r. The modulation induces frequencies in the cavity of $\omega_n = \omega_o \pm n\omega_M$ with n integer and

$$\exp(j\omega_n T_r) = \exp(-jn\delta) \tag{7.4.21}$$

(7.4.17) also holds for $\omega = \omega_n$, provided that any modulation is negligible and the amplitudes A_n cancel throughout the equation. Writing $|\tau_L/T_r| = \varepsilon$, and assuming that δ is small, use a Taylor expansion for $1/[r+\exp(-jn\delta)]$ and equate gain and loss terms in (7.4.17), to yield the required value of $1/\tau_g$ for the frequency ω_n

$$(\tau_L/\tau_{gn}) = 2\varepsilon jn(2\pi-\delta) + [(1-r)/(1+r)] - [j2rn\delta/(1+r)^2] + [r(n\delta)^2(1-r)/(1+r)^3] \tag{7.4.22}$$

The same phase of all the modes at the reference time $t=0$ can be obtained if the gain term has a phase that is independent of frequency, i.e. if the imaginary terms in (7.4.22) vanish, giving

$$\delta = 2\pi\varepsilon(1+r)^2/r + \text{terms in } \varepsilon^2 \tag{7.4.23}$$

The term ε may easily be $<10^{-3}$ so that the assumption that δ is small is valid. The interaction between the cavity and the active medium is seen to 'pull' the rate of change of phase and so alter the frequency by a small amount. The required gain at frequency ω_n is then given from (7.4.22) and (7.4.23) as

$$(\tau_L/\tau_{gn}) = \alpha + (\beta' n)^2 \tag{7.4.24}$$

where

$$\alpha = [(1-r)/(1+r)]; \quad \beta'^2 = [(2\pi\varepsilon)^2(1-r^2)/r]$$

In practical lasers, the gain is a function of frequency and this can be taken into account by writing

$$(\tau_L/\tau_{gn}) = (\tau_L/\tau_{go}) - (\gamma n)^2 \tag{7.4.25}$$

allowing for gain curvature as in section 7.2, but with an appropriately defined value of γ for this problem. (7.4.24) can then be rewritten with the required value of τ_L/τ_{g0} at the nth mode

$$(\tau_L/\tau_{g0})_n = \alpha + (\beta n)^2 \tag{7.4.26}$$

where $\beta^2 = \beta'^2 + \gamma^2$.

One possible effect of an external cavity can be seen immediately. The frequency can be established around ω_o because the gain required at ω_n is higher than that required at ω_o. The cavity is said to possess 'loss curvature'. In the steady state, further frequency stabilisation arises through the rate of change of phase. Cavity stabilisation of a laser is a well-known technique. Composite cavities can be used to enhance the loss curvature and increase the gain stability, with the distributed feed-back laser really being a composite cavity laser where the curvature is so sharp that only a single mode is permitted. In contrast, for mode locking, ideally $\beta' < \gamma$ so that the material, and not the external cavity, limits the optical bandwith which then determines the temporal width of the output pulse.

7.4.4 *Mode locking*

Assume now that the overall gain/loss term is now modulated:

$$[\tau_L/\tau_g] = G_o(1-2M+2M\cos\omega_M t) \tag{7.4.27}$$

This modulation can arise in different ways. First the gain can be altered actively. For example, in a laser diode the current can be changed by changing the electron density so that the gain $g[n(t)-n_o]$ varies with time. Secondly the electron populations can be changed by the photons stimulating the electrons. Periodic optical pumping is used in some mode locked laser systems. Thirdly, the light from the laser itself can depopulate the higher energies and reduce the gain. This gain saturation as the photon intensity increases also modulates the gain. All the effects are assumed to combine to give a change of gain approximately as in (7.4.27). The dynamic effects of gain saturation would need the electron rate equation to be explicitly worked out and would determine the effective value of M to be used.

The modulation of the gain term $(1/\tau_g)$ mixes with the amplitudes a and produces outputs at angular frequencies $\omega_n = \omega_o \pm n\omega_M$. The central frequency is assumed to adjust according to (7.4.18). It is easy to find that any small required adjustment of ω_o automatically occurs through the frequency pulling effects of the external cavity.

Suppose now that the total amplitude of the signal is $\sum\{A_n \exp j\omega_n t\}$. The amplitude A_n of a single mode can no longer be cancelled as it has been in (7.4.22). A modification is required to (7.4.26) to allow for the coupled frequencies. With a little rearrangement, splitting the cosine modulation into its component exponentials leads to:

$$G_oM[-2A_n+A_{n-1}+A_{n+1}] = [\alpha - G_o + (\beta n)^2]A_n \tag{7.4.28}$$

Fortunately $\beta \ll 1$ so that writing $\beta n^2/(G_oM)^{1/2} = x^2$, with $\delta x^2 = \beta/(G_oM)^{1/2}$, (7.2.28) can be turned into an approximate differential equation in x. The gain is adjusted so that

$$[G_o - \alpha]/[\beta(MG_o)^{1/2}] = (2N+1) \tag{7.4.29}$$

For a gaussian pulse it will be found that $N=0$ is required, so with $\beta \ll 1$

$$G_o \sim \alpha \tag{7.4.30}$$

for M close to unity. The finite difference form for $A(x)$ is then

$$(1/\delta x^2)(A_{\delta x}+A_{-\delta x}-2A)+(2N+1-x^2)A=0 \tag{7.4.31}$$

(7.4.31) is the finite difference approximation to

$$d^2A/dx^2+(2N+1-x^2)A=0 \tag{7.4.32}$$

The significance of a general N is considered later, but here, with zero spontaneous emission, $N=0$ is appropriate and gives a gaussian solution $\exp(-x^2/2)$, which when converted back into integer form is

$$A_n \approx A_o \exp[-\beta n^2/2(G_o M)^{1/2}] \tag{7.4.33}$$

Comparing (7.4.33) with (7.4.3), the gaussian pulse has the 'modal variance' $M_o^2 = 2\pi(MG_o)^{1/2}/\beta$. The phases are all locked together by the modulation so that, as for (7.4.3), the optical intensity is a periodic pulse with an approximate envelope $\exp[-2\pi(M_o t/T_r)^2]$, and a period $2\pi/\omega_M \sim T_r$.

The FWHM width of the pulse is determined by the combination of bandwidth of the material gain along with the system bandwidth, imposed by the interaction time and reflector. If the system bandwidth were the only relevant limit then

$$T_{fwhm} = [T_r/M_o][2(\ln 2)/\pi]^{1/2} \tag{7.4.34}$$

Substituting for the normalised variables

$$T_{fwhm} = F[T_r(\tau_L \tau_{go})^{1/2}]^{1/2} \tag{7.4.35}$$

where $F = [(\ln 2)/2\pi]^{1/2}[(1-r^2)/(Mr)]^{1/4}$ and the value of F (noting the fourth root of the function of reflectivities) varies around 0.6 except at extreme values of r.

For a diode laser, around 100 μm long with untreated cleaved facets, one finds that

$$\tau_L \sim \tau_{go} \sim 1 \text{ ps} \quad \text{and} \quad T_r \sim 1 \text{ ns} \rightarrow T_{fwhm} \sim 10 \text{ ps}$$

There have been a few reported results of shorter pulses generated from semiconductor lasers, but the analysis indicates some of the difficulties. The discussion below considers the problems further, along with possible solutions.

7.4.5 *Discussion*

The analysis for a coupled external cavity with a single laser resonance shows that as the coupling becomes tighter so τ_L becomes shorter and shorter, and in principle the pulse width can become limited by the bandwidth of the lasing material. (If there is dispersion in the system so that different frequencies travel with different velocities then again there is a limit but this is not discussed here.) Antireflection coating to the laser facet increases the coupling and so shortens the coupling time constant and the overall limitations of the multiple reflections of the external cavity. The effects of multiple reflections in the external cavity can also be removed by having the interface at an angle so that any reflected light is scattered out of the resonant cavity (see, for example, the schematic angling of the laser facets in Fig. 7.4(c)). While this may lose optical energy and require a higher gain from the laser, the laser and coupled cavity are effectively a single resonator unit. Slight angling of all facets is an effective method for removing a subcavity potentially formed by any set of parallel reflecting faces.

The problem of subcavities are particularly significant for diode lasers which usually do not have facet angling or antireflection coating so that there are usually several longitudinal resonances, ω_{i1}, $\omega_{i2}, \ldots$, internally within the laser. The separation ω_{ir} of these internal modes depends on the optical round-trip within the laser diode. The system bandwidth of the composite cavity can be narrower than ω_{ir}. Each of the internal modes is then separately modulated by the gain to give a periodic amplitude:

$$\mathscr{A}(t) = \sum_p \{A_p A(t) \exp j(\omega_{ip} t + \phi_p)\} \qquad (7.4.36)$$

If the phases of the internal resonances were locked, then Fourier analysis would indicate that $\mathscr{A}(t)$ is a train of short pulses, with a round-trip time of the laser diode cavity, modulated by the longer pulse $A(t)$ with the round-trip time of the external cavity. If the phases are not locked then the pulse $A(t)$ modulates noise with certain enhanced frequencies.

In systems where the bandwidth of the material rather than the system is the limiting factor, then $\beta = \gamma$ in (7.4.26). The limit again depends on tight coupling of the lasing material to the optical cavity. The final limitation must arise through (7.4.4) and the material gain bandwidth. Certain types of dye lasers use dyes with large optical bandwidths, and pulses shorter than 100 fs have been produced.

This brief discussion does not do justice to the difficulties of getting short pulses from mode locked systems. The various problems that arise are a specialist study, but again rate equations help to understand many of these features.

7.4.5 *Spontaneous emission*

It was seen for normal laser action that spontaneous emission could excite additional laser modes in multimode systems. So, here, spontaneous emission can excite other patterns of mode locked laser pulses. If spontaneous emission had been retained then the form of (7.4.32) should have been

$$(d^2A/dx^2)+(1+\theta-x^2)A=S \tag{7.4.37}$$

where S is the excitation from spontaneous emission and $\theta \ll 1$ so that the gain has been set nearly at the value for $N=0$.

However, one can define $A_N(x)$ satisfying

$$d^2A_N/dx^2+(2N+1-x^2)A_N=0 \tag{7.4.38}$$

where

$$A_N = H_N(x) \exp(-\tfrac{1}{2}x^2)$$

and $H(x)_N$ are Hermite polynomials. The set of A_N form an orthogonal set of functions so that

$$\int_{+\infty}^{+\infty} A_N A_M \, dx=0 \tag{7.4.39}$$

giving

$$S(t)=\sum\{S_N H_N(x) \exp(-\tfrac{1}{2}x^2)\} \tag{7.4.40}$$

where

$$S_N = \int_{-\infty}^{+\infty} S(x)A_N(x)\, dx \Big/ \int_{-\infty}^{+\infty} [A_N(x)]^2 \, dX$$

The solution to (7.4.37) is then given from

$$A(t)=\sum\{a_N A_N(x)\} \tag{7.4.41}$$

with

$$a_N=-S_N/(2N-\theta)$$

The zero-order mode can be excited to the greatest extent by keeping $\theta \ll 1$ (i.e. adjusting the gain): $a_0 \sim S_0/\theta$.

Just as spontaneous emission excited several longitudinal modes in a steady state laser operation, so, here, the spontaneous emission excites new patterns of pulses with temporal substructures different from the clean gaussian pulse, appropriate to zero spontaneous emission. This set A_N of temporal modes excited by the spontaneous emission has been called 'supermodes'. The periodic short burst of output energy is then combinations of the temporal supermodes and possibly transverse spatial modes of the optical cavity. The latter are not considered here.

7.5 Photon rate equations from Maxwell's equations

7.5.1 *Maxwell's equations with electrically active media*

As a final topic, rate equations are developed from Maxwell's equations to show how quantum and classical concepts tie together. The work will show the relation of material properties to the rate equations and demonstrate the importance of stored energy in oscillating systems. The reader needs a knowledge of Maxwell's equations,[74,75] but a brief recapitulation is given of some relevant equations to ensure that terminology is defined.

In a material, an applied electrical field can polarise the electron cloud around the nucleus of each atom to give an average dipole strength/unit volume

$$\boldsymbol{P}=\chi\varepsilon_0\boldsymbol{E} \tag{7.5.1}$$

(In the most general form χ is a polarisability tensor with $\boldsymbol{P}$ and $\boldsymbol{E}$ not necessarily in the same direction, but here χ is taken to be a scalar.) The quantum theory for the polarisability of atoms shows that valence electrons changing between quantum states of energy $\mathscr{E}_1$ and $\mathscr{E}_2$ can, given appropriate symmetry for the electron clouds in these states, create an electric dipole which responds strongly to electric fields changing at the (angular) frequency $\omega_{12}=(\mathscr{E}_1-\mathscr{E}_2)/\hbar$. In response to fields varying generally as $\exp j\omega t$, χ will be complex:

$$\chi=\chi'+j\chi'' \tag{7.5.2}$$

The significance of the real and imaginary parts of χ can be seen by noting that $\partial\boldsymbol{P}/\partial t$ is equivalent to a current density $\boldsymbol{J}_p$. If $\chi=\chi'$ is real, then the polarisation current density $\boldsymbol{J}_p$ cannot do any work averaged over time (see problem 7.3). The electric displacement vector is then

$$\boldsymbol{D}=\varepsilon\boldsymbol{E}=\boldsymbol{P}+\varepsilon_0\boldsymbol{E};\quad \varepsilon=(1+\chi')\varepsilon_0 \tag{7.5.3}$$

If at the frequency ω one has $\chi=j\chi''$, then there is a polarisation current density

$$\boldsymbol{J}_p=-(\omega\chi''\varepsilon_0)\boldsymbol{E} \tag{7.5.4}$$

The negative sign shows that $\boldsymbol{J}_{\mathrm{p}}$ is in antiphase with the electric field $\boldsymbol{E}$ and so gives *out* power if $\chi''>0$. If $\chi''<0$ then power is absorbed as in any 'lossy' material. The total displacement current density is then given by

$$\partial\boldsymbol{D}/\partial t=-\omega\chi''\varepsilon_0\boldsymbol{E}+\varepsilon\,\partial\boldsymbol{E}/\partial t-\boldsymbol{J}_{\mathrm{s}} \tag{7.5.5}$$

where ε is given by (7.5.3) and $\boldsymbol{J}_{\mathrm{s}}$ models the spontaneous emissions by giving fluctuations in the polarisation current. For most materials, χ'' is negative corresponding to stimulated absorption, but with enough carriers in appropriate higher energy levels and values of $\omega\sim\omega_{12}$, χ'' can become positive corresponding to stimulated emission.

Maxwell's equations for such a medium with fields varying around the frequency ω can then be recast as

$$\partial\boldsymbol{e}/\partial t=c_{\mathrm{m}}\,\mathrm{curl}\,\boldsymbol{h}+g\boldsymbol{e}+\boldsymbol{i}_{\mathrm{s}} \tag{7.5.6}$$

$$\partial\boldsymbol{h}/\partial t=-c_{\mathrm{m}}\,\mathrm{curl}\,\boldsymbol{e} \tag{7.5.7}$$

where $e=(\varepsilon/\mu)^{1/4}E$; $h=(\mu/\varepsilon)^{1/4}H$; $c_m=1/(\mu\varepsilon)^{1/2}$; $g=\omega\chi''/(1+\chi')$; $i_s=c_m(\mu/\varepsilon)^{1/4}J_s$; $\varepsilon=(1+\chi')\varepsilon_0$.

By assuming that the fields vary as $e=E(t)\exp j\omega t$ with the real amplitude varying slowly compared to ω, the stored energy in a volume Φ is given by:

$$\mathscr{E}=(1/4c_m)\iiint_\Phi (e^*\cdot e+h^*\cdot h)\,dV \tag{7.5.8}$$

The electromagnetic power flow F out of a closed surface Σ is given from the complex Poynting vector integrated over the surface:

$$F=\tfrac{1}{4}\iint_\Sigma (e\times h^*+e^*\times h)\cdot n\,dS \tag{7.5.9}$$

where n is the unit outward normal of the surface. For fields varying as $\exp j\omega t$, (7.5.8) and (7.5.9) have removed the rapid variations, at frequencies 2ω, of instantaneous energy and power by forming products such as $e^*\cdot e$ rather than $e\cdot e$. But growth and decay are still observed in F and ε averaging over a cycle. (Factors of $\frac{1}{4}$ instead of $\frac{1}{2}$ in (7.5.8) and (7.5.9) recognise that the average values of $(A\cos\omega t)^2/2$ is $\frac{1}{4}A^2$.)

Two standard vector theorems are used frequently:

(i) For any two vectors e and h

$$\operatorname{div}(e\times h)=h\cdot\operatorname{curl} e-e\cdot\operatorname{curl} h \tag{7.5.10}$$

(ii) $$\iiint_\Phi \operatorname{div}(e\times h)\,dV=\iint_\Sigma (e\times h)\cdot n\,dS \tag{7.5.11}$$

where the volume Φ is surrounded by a surface Σ with unit outward normal n.

7.5.2 *The energy rate equation*

Form from (7.5.7) and (7.5.6) the equation

$$\begin{aligned}\operatorname{div}(e\times h^*)&=(h^*\cdot\operatorname{curl} e-e\cdot\operatorname{curl} h^*)\\&=-(1/c_m)(h^*\cdot\partial h/\partial t+e\cdot\partial e^*/\partial t)+(g/c_m)e^*\cdot e\\&\quad+(i_s^*\cdot e/c_m)\end{aligned} \tag{7.5.12}$$

Integrating over the interaction region Φ with the emitting surface(s) denoted by Σ (unit outward normal n) yields the complex power out:

$$F+jF_r=\tfrac{1}{2}\iint_\Sigma (e\times h^*)\cdot n\,dS=\tfrac{1}{2}\iiint_\Phi \operatorname{div}(e\times h^*)\,dV \tag{7.5.13}$$

Assuming that power flows out of the system only along the z-axis with unit direction $\boldsymbol{n}$,

$$F+jF_r=\tfrac{1}{2}\iiint_{\Phi}(\partial/\partial z)[(\boldsymbol{e}\times\boldsymbol{h}^*)\cdot\boldsymbol{n}]\,\mathrm{d}V \qquad (7.5.14)$$

with the transverse integrations giving zero net contribution.

Now add (7.5.12) to its own complex conjugate and integrate over the interaction volume as in (7.5.13) to form the power equation:

$$F=-\partial\mathscr{E}/\partial t+g[\mathscr{E}-\delta\mathscr{E}]+P_s \qquad (7.5.15a)$$

where

$$\delta\mathscr{E}=(1/4c_m)\iiint_{\Phi}(\boldsymbol{h}^*\cdot\boldsymbol{h}-\boldsymbol{e}^*\cdot\boldsymbol{e})\,\mathrm{d}V \qquad (7.5.15b)$$

$$P_s=(1/4c_m)\iiint_{\Phi}(\boldsymbol{i}_s\cdot\boldsymbol{e}^*+\boldsymbol{i}_s^*\cdot\boldsymbol{e})\,\mathrm{d}V \qquad (7.5.15c)$$

(7.5.15) can be shown to be equivalent to the usual rate equation for photons. It is known that the average stored magnetic energy equals the average stored electric energy at a resonance (see next subsection) so that $\delta\mathscr{E}$ is zero. The spontaneous emission P_s which increases the electric energy is only that part which couples with the electric field at the frequency ω under consideration. Spontaneous emission over a broad spectrum of frequencies can be modelled as a set of pulses of current at random points and random times, and the coupling to the actual electric field is carried out through an integration such as in (7.5.15c).

The value of the power flow F out of the interaction region gives the physical interpretation of the escaping photon energy loss. It can be seen also to give the correct result by considering (7.5.14) and its complex conjugate. In the absence of any growth of power:

$$(\boldsymbol{e}\times\boldsymbol{h}^*+\boldsymbol{e}^*\times\boldsymbol{h})\cdot\boldsymbol{n}=\text{power flow forward}-\text{power flow back} \qquad (7.5.16)$$

With an active medium, the forward components of e and h each grow as $\exp \alpha z$, while the reverse components grow in the reverse direction and so vary as $\exp -\alpha z$. Hence, differentiation with z changes a sign to give

$$\begin{aligned}(\partial/\partial z)(\boldsymbol{e}\times\boldsymbol{h}^*+\boldsymbol{e}^*\times\boldsymbol{h})\cdot\boldsymbol{n}\\ &=2\alpha\ (\text{power flow forward}+\text{power flow back})\\ &=2\alpha\ (\text{energy density}\times\text{group velocity in material})\\ &=2\alpha c_{gm}\mathscr{E}/\Phi\end{aligned} \qquad (7.5.17)$$

Now integrate (7.5.17) over the interaction volume, Φ, and make use of the spatial growth factor 2α for power found in section 6.2.5:

$$2\alpha c_{gm}=G(n_t-n_o)=G_t=1/\tau_p \qquad (7.5.18)$$

Hence, the flux F giving the rate of escape of energy is

$$F = \mathscr{E}/\tau_p \tag{7.5.19}$$

The photon density P is defined from the energy in the interaction volume $\mathscr{E} = P\Phi\hbar\omega$. The coupled spontaneous emission P_s defines β from $P_s = \hbar\omega\Phi\beta n_s/\tau_r$ to recover the rate equation (6.4.1):

$$P/\tau_p = -(\partial P/\partial t) + gP + (\beta n_s/\tau_r) \tag{7.5.20}$$

It follows that the material gain $G(n - n_o)$ used previously is the value of $g = \omega\chi''/(1+\chi')$ averaged over the interaction region. The terms in the normal photon rate equation are all then identified.

7.5.3 *Equality of electric and magnetic energy*

With steady state time variations of the fields as $\exp j\omega t$, integrating (7.5.12) over Φ yields the complex power flow

$$F + jF_r = -2j\omega\delta\mathscr{E} + g(\mathscr{E} - \delta\mathscr{E}) + P_s + jP_{sr} \tag{7.5.21}$$

where

$$jP_{sr} = (1/4c_m)\iiint (\boldsymbol{i}_s \cdot \boldsymbol{e}^* - \boldsymbol{i}_s^* \cdot \boldsymbol{e})\,\mathrm{d}V$$

The imaginary part of the complex power is well known in conventional a.c. circuits to give the magnitude of the reactive power where energy flows back and forth at a frequency 2ω between energy storage components (inductors and capacitors). The same result holds in field theory, but for the isolated laser with no external feedback there is no external storage medium which can reflect back power. The reactive power at the facets of such a matched laser must be zero:

$$F_r = 0 = -2\omega\delta\mathscr{E} + P_{sr} \tag{7.5.22}$$

In the absence of any spontaneous emission, $\delta\mathscr{E} = 0$, so that average magnetic and electric energies are equal. In the presence of spontaneous emission there is fluctuation between the two energies. The phase of any spontaneous emission must be entirely random so that the magnitude of both P_s and P_{sr} are expected to be equal. P_s can be estimated from the recombination and spontaneous coupling coefficient as in (7.5.20). At very low light levels, the quantum fluctuations in $\boldsymbol{E}$ and $\boldsymbol{H}$ will add further difficulties, but the light output is well above this level and the fluctuations are modelled by spontaneous emission $P_s + jP_{sr}$. The fluctuation in $\delta\mathscr{E}$ will be significant for frequency stabilisation.

7.5.4 *Slater's perturbation theorem*

The importance of changes of stored energy in stabilising the frequency is well known in radio and microwave devices and applies equally to optical devices. We derive here a significant theorem, known in microwaves as Slater's perturbation theorem,[75] giving the rate of change of frequency with energy.

Consider some active device in which a perturbation is made to a local volume $\delta\Phi$ with surface $\delta\Sigma$ changing the reactive power flowing between $\delta\Phi$ and the main volume Φ. From (7.5.21), the change in reactive power flowing 'out' from $\delta\Phi$ is

$$\delta F_{\mathrm{r}} = -2\omega\,\delta\mathscr{E} = -(\omega/2c_{\mathrm{m}})\iiint_{\delta\Phi}(\boldsymbol{h}^*\cdot\boldsymbol{h} - \boldsymbol{e}^*\cdot\boldsymbol{e})\,\mathrm{d}V \quad (7.5.23)$$

The result of this change is to alter the fields and the frequency of emission so that in the main interaction volume there are changes $\delta\boldsymbol{e}$, $\delta\boldsymbol{h}$, $\delta\omega$ which restore the balance of reactive power flow to zero and restore the equality of electric and magnetic energy within the active system. For such changes in the main volume, there must be a change in reactive power flow 'out' from Φ.

The correct identification of this change of reactive power can prove a problem when tackled directly from field theory. However, in circuit theory the reactive power flow is given by $\frac{1}{2}I^*IX$ so that for a change in the reactive element one looks for $\frac{1}{2}I^*I\delta X$, as the change in reactive power. However, considering a few standard relationships:

$$V = jXI; \quad V^* = -jXI^*$$
$$\delta VI^* = (jI\delta X + jX\delta I)I^* = jI\,\delta XI^* - \delta IV^*$$
$$\tfrac{1}{2}jI^*I\delta X = \tfrac{1}{2}(\delta VI^* + V^*\delta I)$$

Fig. 7.5. Notation for Slater's perturbation theorem. Total reactive power F_{r} out is zero. Perturbation of electric and/or magnetic fields changes frequency.

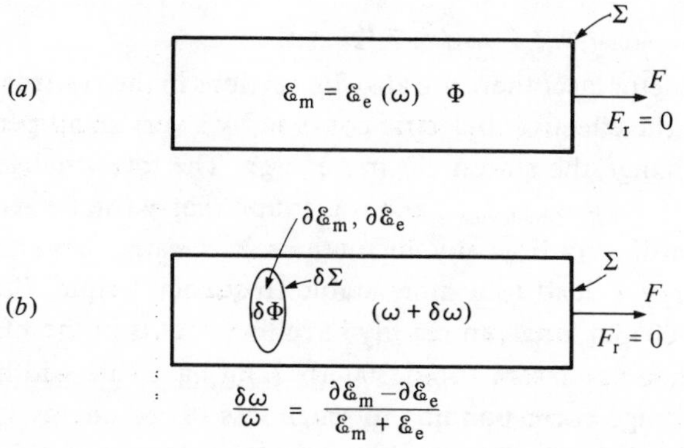

Interpreting this in terms of reactive power flow in electromagnetic theory:

$$j\delta F_r = \tfrac{1}{4}\iint_{\Sigma \neq \delta\Sigma} (\delta \boldsymbol{e} \times \boldsymbol{h}^* + \boldsymbol{e}^* \times \delta \boldsymbol{h}) \cdot \boldsymbol{n}\, \mathrm{d}S$$
$$- \text{complex conjugate} \qquad (7.5.24)$$

The reactive power flowing at the ports other than $\delta\Sigma$ is still zero for a matched device so that (7.5.23) has to equal (7.5.24), and it is this balance of reactive power that determines the change in frequency.

Considering only the reactive elements and neglecting the stimulated polarisation current $\boldsymbol{J}_p$ which does not contribute to changes of the reactive power:

$$j\omega\, \delta \boldsymbol{e} + j\, \delta\omega \boldsymbol{e} = c_m \operatorname{curl} \delta \boldsymbol{h} \qquad (7.5.25)$$

$$-j\omega\, \delta \boldsymbol{h} - j\, \delta\omega \boldsymbol{h} = c_m \operatorname{curl} \delta \boldsymbol{e} \qquad (7.5.26)$$

Using the standard relationships of (7.5.10) and (7.5.11) enables (7.5.24) to be rewritten as a volume integral, and after a little rearrangement:

$$\delta F_r = (\delta\omega/2c_m) \iiint_{\Phi} (\boldsymbol{e}^* \cdot \boldsymbol{e} + \boldsymbol{h}^* \cdot \boldsymbol{h})\, \mathrm{d}V = 2\delta\omega\mathscr{E} \qquad (7.5.28)$$

The changes $\delta\boldsymbol{e}$ and $\delta\boldsymbol{h}$ are found to cancel in the volume integral leaving the first-order change in reactive power flow entirely in terms of the first-order change in frequency. Matching up the two changes of reactive power flow (caused by the changed volume $\delta\Phi$ and the changed frequency $\delta\omega$):

$$\delta\omega/\omega = (\delta\mathscr{E}_e - \delta\mathscr{E}_m)/\mathscr{E} \qquad (7.5.29)$$

where $\delta\mathscr{E}_e$ and $\delta\mathscr{E}_m$ are the changes in electric and magnetic energy (negative if energy is removed) caused by the perturbation $\delta\Phi$.

From the discussion leading to (7.5.22) it follows that, for all types of laser, spontaneous emission creates a fluctuation in $(\delta\mathscr{E}_e - \delta\mathscr{E}_m) = \mathscr{E}_{noise}$ which will lead to fluctuations in $\delta\omega$:

$$\delta\omega_{spontaneous}/\omega = \mathscr{E}_{noise}/\mathscr{E} = P_s/2\omega\mathscr{E} \qquad (7.5.30)$$

In a semiconductor laser there are also fluctuations in the electron density which change the effective dielectric constant by a very small percentage and so also change the stored electric energy. The total change in line width is then $(1+\gamma)\delta\omega_{spontaneous}$ and it is found that γ can be around 25. However, regardless of how any fluctuations are created, larger amounts of stored energy $\mathscr{E}$ lead to a more stable frequency output. Stable gas lasers about 300 mm long can readily have line widths in the kHz range while semiconductor lasers about 300 μm long have line widths in the tens of MHz range corresponding to much less stored energy (problem 7.6) and the additional effects of fluctuations in electron density.

In order to make semiconductor lasers have narrower line widths (approximating better to a single frequency), diffraction gratings have been built inside the devices (distributed feedback lasers, Fig. 7.6). One can either consider the diffraction grating as a frequency selective filter within the device or alternatively regard the grating as a method of introducing multiple reflections which increase the stored energy for a given power flow, thereby increasing stability. Dramatic reductions of line width into the kHz range have been made giving useful lasers for coherent optical communications, especially with external cavities.

7.5.5 *Amplitude rate equations*

To develop a formalism that permits one to see some of the approximations that arise in an amplitude rate equation, as used for mode locking in section 7.4, go back to (7.5.6) and (7.5.7). Maxwell's divergence relationships require div $\boldsymbol{D}=0$ so that, assuming an approximately uniform χ, div $\boldsymbol{e}=0$, giving

$$\partial(\text{curl } \boldsymbol{h})/\partial t = -c_{\text{m}}[(k_z + j\alpha)^2 + k_{\text{t}}^2]\boldsymbol{e} \tag{7.5.31}$$

Here ∇^2 is replaced by a complex constant assuming constant uniform propagation constants in the active device, with k_{t} the transverse value of the propagation, k_z the longitudinal propagation and α the spatial growth constant. The important point to emphasise here is that these k-values are fixed by the spatial boundary conditions for the electric and

Fig. 7.6. Schematic diagram of distributed feedback laser diodes (DFBs). Layers 1, 2 and 3 are designed as in a conventional heterojunction laser to confine the electrons, holes and photons mainly to layer 2 where strong stimulated emission occurs. The substrate is rippled to provide a strong feedback (reflections) to the waves leading out into layer 3. Reflections occur at certain wavelengths $\lambda = 2d/n$ (n typically designed to be 1 to 3). These reflections provide a high energy storage within the diode for a given power flow.

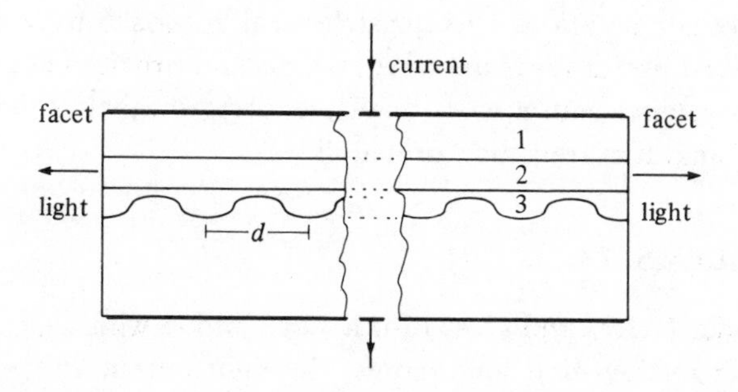

magnetic fields. The solution of electromagnetic boundary value problems with the resultant modes must be left for further reading but, for example, in an ideal Fabry–Perot of length L, it was required that $k_z L = 2n\pi$. The transverse geometry similarly fixes the effective value of k_t, while the terminations fix the value of α as in section 6.2.5. The geometry thus fixes the values of k_z, k_t and α. An angular frequency ω_o can be defined by

$$k_z^2 + k_t^2 = (\omega_o / c_m)^2 \tag{7.5.32}$$

so for small values of gain and transverse propagation constant

$$\partial(\text{curl}\, \boldsymbol{h})/\partial t = -(\omega_o^2/c_m)\boldsymbol{e} - j2\alpha\omega_o \boldsymbol{e} \tag{7.5.33}$$

Define a complex vector amplitude $\boldsymbol{a} = [(\omega_o \boldsymbol{e}/c_m) - j\,\text{curl}\,\boldsymbol{h}]$ and combine (7.5.33) and (7.5.6) to give

$$\partial \boldsymbol{a}/\partial t = j\omega_o \boldsymbol{a} - (\alpha c_m)(\boldsymbol{a} + \boldsymbol{a}^*) + (g/2)(\boldsymbol{a} + \boldsymbol{a}^*) + (\omega/c_m)\boldsymbol{i}_s \tag{7.5.34}$$

It can be seen that $\boldsymbol{a}$ varies approximately as $\exp j\omega_o t$ so that $\boldsymbol{a}^*$ varying as $\exp(-j\omega t)$ does not couple well into (7.5.34). This 'rotating wave approximation' leaves the result:

$$\partial \boldsymbol{a}/\partial t = j\omega_o \boldsymbol{a} - (\boldsymbol{a}/2\tau_p) + (g/2)\boldsymbol{a} + (\omega/c_m)\boldsymbol{i}_s \tag{7.5.35}$$

where section 6.2.5 is used (as in (7.5.17)) to evaluate α in terms of τ_p.

This equation gives the amplitude rate equation assuming a uniform interaction over the volume with g identified as the gain of the medium. The amplitude can be scaled by some factor K so that integrating over the interaction volume:

$$\iiint_\Phi (K\boldsymbol{a})^* \cdot (K\boldsymbol{a})(\mathrm{d}V/\Phi) = P$$

where P is the total photon number. Then dropping explicit writing of the scale factor K, $\boldsymbol{a}^* \cdot \boldsymbol{a} = P$ as used in (7.4.5). The extra terms in (7.4.6) arise from the external coupling in a straightforward way. The scalar a in (7.4.6) simply states that the mode is concerned with $\boldsymbol{a}$ in one direction (single polarisation of the electric field) so that there is no ambiguity in dropping the vector notation. This quasiclassical approach to quantum theory has proved useful for many laser systems and provides engineers and applied physicists with a more readily understood theory than the full rigours of quantum transition probabilities.

PROBLEMS 7

7.1 Consider (7.3.2) and (7.3.3) for a *single* mode with a small but non-zero value of β and express the light output $P\Phi/\tau_p$ as a function of the drive level. Show that $I_t = e\Phi n_t/\tau_r$ gives approximately the 'threshold' current where $n_t = n_o + 1/G\tau_p$. Estimate the light output for laser (i) when $I = I_t$ with spontaneous emission, (ii) as a function of drive with $I \gg I_t$.

7.2 A problem may be found to convince the reader/student that the general 'inspection' methods of rate equations give correct answers. For the steady state in section 7.4.2, show that the value of $b_L/b_R = [(1/\tau_L) + G - (1/\tau_R) - j2\,\delta\omega]/[(1/\tau_L) + (1/\tau_R) - G + j2\,\delta\omega]$ where $\delta\omega = \omega - \omega_o$. Neglect s in (7.4.6).

Contrast this with a reflection coefficient of a shunt $LC(-R_g)$ circuit fed at the left hand end by a transmission line of impedance R_L. Begin by showing that the admittance of the resonant circuit close to resonance ω_o is $Y = -(1/R_g) + 2jC\,\delta\omega$. Take $1/R_R$ as the admittance shunted across the circuit at the right hand end. The resistance $-R_g$ is negative, corresponding to gain in the medium forming the resonant circuit. Find the overall reflection coefficient of the resonant circuit with negative resistance and load R_R.

What is the condition for oscillation (lasing) neglecting any spontaneous emission?

7.3 Given a real polarisability, (7.5.2), show that the rate of energy transfer from the polarisation current density $\boldsymbol{J}_p$ is proportional to $\int \chi'\varepsilon_o \boldsymbol{E}\cdot\partial\boldsymbol{E}/\partial t\,\mathrm{d}t$. Explain why, over a sufficiently long time, this current cannot give any net power.

7.4 An argument of time reversal was used in chapter 5 to good effect. Suppose in section 7.4.2 the active region contains P_o photons when the gain/loss is suddenly switched to give $-G + (1/\tau_R) = 0$. The rate of photons feeding out is $|b_L^2|$, and this gives a decay in the stored photons according to $\exp(-t/\tau_L)$. Now reverse time so that the same power $|b_L^2|$ is fed into the active region through the amplitude $b_R = (1/\tau_L)^{1/2}a$.

What is the growth of stored photons? Show that with a coupling factor k from b_R to a, keeping G as above,

$$\mathrm{d}a/\mathrm{d}t = [j\omega_o - (1/2\tau_L)]a + k(1/\tau_L)^{1/2}a$$

Hence evaluate k.

7.5 In evaluating (7.5.13), the power escaping sideways was ignored. Splitting ∇ into $\nabla_t + (\partial/\partial z)\hat{z}$, show that a non-zero integration of the terms $\iiint_\Phi \nabla_t\cdot(\boldsymbol{e}\times\boldsymbol{h}^* + \boldsymbol{e}^*\times\boldsymbol{h})\,\mathrm{d}V$ must lead to an additional term $-P/\tau_{pi}$ in the rate equation representing photons being absorbed at the edges of the waveguide (i.e. photons escaping sideways).

7.6 If the phase of the spontaneous dipole moment is arbitrary, then the in phase and reactive dipole spontaneous moments are equal so that in (7.5.21) $P_{sr} = P_s = \beta n_t \hbar\omega\Phi/\tau_r$. Use (7.5.22) and (7.5.30) to show that one expects a frequency fluctuation (line width) for a single mode injection laser of the order of

$$\delta\omega/\omega \sim \beta \ln(1/R)/(\eta 4N\pi)$$

where β is the spontaneous coupling coefficient, η is the laser efficiency, N is the number of wavelengths in the laser cavity and R is the facet reflectivity. (Neglect any effects caused by fluctuations in the electron drive changing the electron density and so changing the effective permittivity of the lasing material - though in practice this may increase the line width by an order of magnitude.[83])

APPENDIX A

Counting states

A.1 Concept of a state: photons

In quantum theory, a state is a condition of a system which is 'stable', so if a measurement identifies a system to be in an ideal state and the measurement is repeated immediately then the same state will be observed.

An example may be found in electromagnetic theory where, once the resonant energy in a lossless LC circuit is identified as giving a frequency $\omega = 1/(LC)^{1/2}$, then the circuit remains at that frequency and this identifies a photon state or mode.

In a Fabry–Perot (one-dimensional plane wave) resonator the modes can be identified by making the system lossless with ideal reflectors forcing the electric field to be zero at both ends ($z=0$ and $z=L$). Then it is possible to support only waves

$$E = E_0 \sin kz, \quad \text{where } kL = n\pi \tag{A.1.1}$$

with n integer so that E vanishes at both ends.

This system actually *shares* two states because if a group of photons were to be measured they could be travelling either to the right or the left or indeed in a mixture. However, if the measurement is repeated immediately then only those photons moving wholly in one or other direction will still be travelling in the same direction, the requirement of an ideal state. From A.1.1, each resonance sharing two states requires $\Delta k = \pi/L$. Hence, each quantum state requires a range of k values

$$\delta k = 2\pi/L$$

Between a change of photon momenta p to $p+\delta p$ ($p=\hbar k$) there are then

$$N_Q = \delta p L/(2\pi\hbar) \tag{A.1.2}$$

quantum states for a resonator of length L.

The same arguments apply to a three-dimensional ideal resonator $L_x \times L_y \times L_z$ with lossless boundaries. Each quantum state requires

$$\delta k_q = 2\pi / L_q; \quad q = x, y, z \tag{A.1.3}$$

Thus between a range of photon momenta $\delta p_x\, \delta p_y\, \delta p_z = \delta V_p$ there are

$$N_Q = V \delta V_p / (2\pi \hbar)^3$$

where $V = L_x L_y L_z$.

This is not quite right because the electric field vector in plane waves can be in two orthogonal directions. If light passes through a polariser such that the E field is vertical, then in an immediate subsequent measurement, using a similarly oriented polariser, the light is found to pass showing that the polarisation has not altered. Consequently polarisation is also a quantum state. The horizontal polarisation is orthogonal to the vertical polarisation showing that there are two independent polarisation states for each plane wave. The final result is therefore

$$N_Q = V \delta V_p / [4(\hbar \pi)^3] \tag{A.1.4}$$

This result shows that states are proportional to the volume of the material and the 'volume' in momentum space occupied by the quantum particles. This result must be translated into terms of energy.

A.2 Electron states in a crystal

Quantum states around an isolated atom have to be found by solving Schrodinger's equation, but, for the nearly free electrons in a crystal, similar arguments to those for photons can be used to define the quantum states; however here it is more usual to argue that the crystal is perfectly periodic. One is therefore concerned with periodic conditions where after a distance L_x (containing many unit cells of the crystal) the system repeats identically. The quantum wave $\psi = \psi_0 \exp jk_x x$ describing the motion in the x-direction must then be periodic leading to

$$k_x L_x = 2n\pi, \quad n \text{ integer}$$

or

$$\delta k_x = 2\pi / L_x \tag{A.2.1}$$

and similarly for y and z, exactly as for (A.1.3). Because for all quantum particles the momentum is related to the k vector: $p = \hbar k$. Moreover, just as photons have two independent polarisations per momentum value, so electrons have two independent 'spins' per momentum value leading to precisely the same result as (A.1.4) for quantum states per δV_p and volume V.

A.3 Density of states

The momentum density of states for both electrons and photons is given by (A.1.3). It is often more useful to have this given in terms of the energy.

For photons and electrons the energy $\mathscr{E}$ is given from $\mathscr{E}=\hbar\omega$.

(i) For photons in a material with phase velocity of light c_{m}

$$(\hbar\omega/c_{\mathrm{m}})^2=(\mathscr{E}/c_{\mathrm{m}})^2=p\cdot p=p^2$$

$$\delta\mathscr{E}=c_{\mathrm{gm}}\,\delta p \tag{A.3.1}$$

where $c_{\mathrm{gm}}=\mathrm{d}\omega/\mathrm{d}k$ is the group velocity, which for many materials is a small percentage different from the phase velocity $c_{\mathrm{m}}=\omega/k$.

(ii) For electrons, with effective mass m^*

$$\hbar\omega=\mathscr{E}=p^2/2m^*$$

$$\delta\mathscr{E}=p\,\mathrm{d}p/m^*=(2\mathscr{E}/m^*)^{1/2}\,\delta p \tag{A.3.2}$$

Between p and $p+\delta p$ there is a shell of momentum space of radius p and thickness δp, so that for both photons and electrons

$$\delta V_{\mathrm{p}}=4\pi p^2\,\delta p$$

The values of p and δp are found in terms of $\mathscr{E}$ and $\delta\mathscr{E}$ from (A.3.2) and (A.3.1), so that from above and (A.1.4) one finds the number of quantum states between energies $\mathscr{E}$ and $\mathscr{E}+\delta\mathscr{E}$ and in a unit volume is given by

(*a*) For electron states:

$$N(\mathscr{E})\delta\mathscr{E}=4\pi(2m^*/h^2)^{3/2}\mathscr{E}^{1/2}\delta\mathscr{E} \tag{A.3.3}$$

From this one can calculate the probable number of electrons in the conduction band, for example, as

$$n=\int_0^\infty \{N(\mathscr{E})/[1+\exp(\mathscr{E}-\mathscr{E}_f)/kT]\}\,\mathrm{d}\mathscr{E} \tag{A.3.4}$$

where $\mathscr{E}$ is measured from the bottom of the conduction band $\mathscr{E}_{\mathrm{c}}$.

(*b*) For photons:

$$N(\mathscr{E})\delta\mathscr{E}=8\pi f^2/(c_{\mathrm{m}}^2c_{\mathrm{gm}})\delta f \tag{A.3.5}$$

Thus the energy in a unit volume at thermal equilibrium (the Planck black body radiation) is given by

$$\mathscr{E}(f)\,\mathrm{d}f=hf\{8\pi f^2/[c_{\mathrm{m}}^2c_{\mathrm{gm}}]/[\exp(hf/kT)-1]\}\,\mathrm{d}f \tag{A.3.6}$$

A.4 Classical states

Counting the 'number' of states of a particle in classical theory is not so clear as counting quantum states for the simple reason that there is a continuum of classical states. The discussion is limited to a gas of ideal classical particles where the energy is

$$\mathscr{E} = \tfrac{1}{2}m(v_x^2 + v_y^2 + v_z^2)$$

In classical theory the state of the particle is described in terms of the momentum

$$\boldsymbol{p} = (mv_x, mv_y, mv_z)$$

so that the density of states at any energy is proportional to the range of momenta that are taken - the volume in momentum space:

$$\mathrm{d}V_\mathrm{p} = \mathrm{d}p_x, \mathrm{d}p_y, \mathrm{d}p_z$$

The Maxwell-Boltzmann distribution becomes

$$n(p_x, p_y, p_z)\,\mathrm{d}p_x\,\mathrm{d}p_y\,\mathrm{d}p_z = A[\exp(-\boldsymbol{p}\cdot\boldsymbol{p}/2mkT)]\,\mathrm{d}p_x\,\mathrm{d}p_y\,\mathrm{d}p_z \tag{A.4.1}$$

Following the discussions for the electrons, there is a shell of momenta of volume $\mathrm{d}V_\mathrm{p} = 4\pi p^2\,\mathrm{d}p$ taking momenta in all directions so that

$$N(p)\,\mathrm{d}p = A[\exp(-p^2/2mkT)](4\pi p^2\,\mathrm{d}p) \tag{A.4.2}$$

The total number of particles N is then

$$N = \int_0^\infty N(p)\,\mathrm{d}p$$

giving $A = N/(2\pi mkT)^{3/2}$.

(Standard integrals of use in Maxwell-Boltzmann calculations are

$$\int_0^\infty \exp(-ax^2)\,\mathrm{d}x = (\pi/4a)^{1/2}; \qquad \int_0^\infty x\exp(-ax^2)\,\mathrm{d}x = 1/2a;$$

$$\int_0^\infty x^2\exp(-ax^2)\,\mathrm{d}x = (\pi/16a^3)^{1/2}) \tag{A.4.3}$$

Hence, the number of particles between $\mathscr{E}$ and $\mathscr{E} + \delta\mathscr{E}$ in such a perfect gas is $N(\mathscr{E})\delta\mathscr{E}$, where

$$N(\mathscr{E}) = [2\pi N/(\pi kT)^{3/2}]\mathscr{E}^{1/2}\exp(-\mathscr{E}/kT) \tag{A.4.4}$$

APPENDIX B

Notes on differences between gas and diode laser rate equations

In a diode laser, the upper energy level $\mathscr{E}_u$ is pumped directly by driving a current. In a gas, doped glass, or similar laser using the energy levels of individual atoms, the electrons are typically allowed to populate $\mathscr{E}_u$ by spontaneous changes from an even higher energy level $\mathscr{E}_U$. Similarly the lower energy level $\mathscr{E}_l$ has the electrons removed from it in a diode by a direct hole current, while in the gas and solid laser the depopulation of the lower level is often accomplished again by spontaneous changes of energy to a yet lower level $\mathscr{E}_L$. The energy $\mathscr{E}_U$ is then filled with electrons by photons pumping energy at $\hbar\omega_{pump} = \mathscr{E}_U - \mathscr{E}_L$, while the emission comes out at $\hbar\omega_{lasing} = \mathscr{E}_u - \mathscr{E}_l$. Fig. B.1 shows a preferred four level scheme.

The details of the rate equations alter, not only through the complexity of the four levels, but also on the details of the probabilities. Suppose the lasing emission occurs by an electron falling between two energy

Fig. B.1.

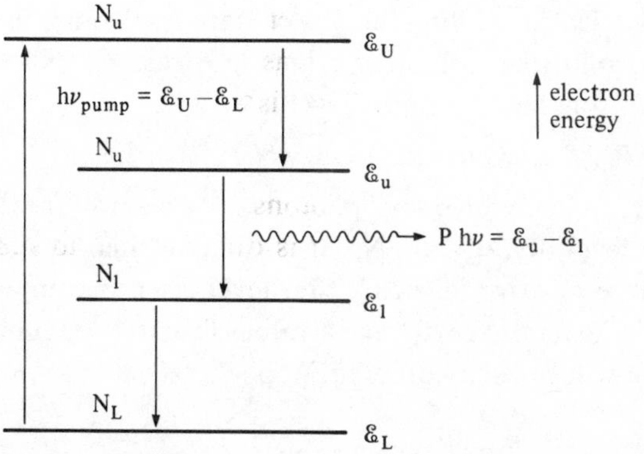

levels $\mathscr{E}_u$ and $\mathscr{E}_l$ with a single state in the atom. Then if there are N_u atoms per unit volume with electrons already in the upper level and no electron in the lower level then the probability of spontaneous emission is just $A_{ul}N_u$, where A_{ul} is the spontaneous emission rate for electrons to fall from $\mathscr{E}_u$ to $\mathscr{E}_1$. Electrons cannot transfer to some other atom so that there is no additional counting of atoms with vacant lower energy sites comparable to the calculations for the semiconductor laser. Similarly the stimulated emission rate is $B_{ul}PN_u$ and the stimulated absorption rate is $B_{lu}PN_l$, where N_l is the number of atoms per unit volume with an electron at $\mathscr{E}_l$ and not an electron at $\mathscr{E}_u$. P is the photon density (Fig. B.1). The photon rate equation is then

$$dP/dT = B_{ul}PN_u - B_{lu}PN_l + \beta A_{ul}N_u - (P/\tau_p) \qquad \text{(B.1.1)}$$

where τ_p is the photon lifetime for the lasing photon density P giving the cavity losses and the useful photon output. The value of β is the coupling coefficient for any spontaneous emission into the lasing mode. As discussed for section 6.2, this is small and the dimensions of a gas laser usually make β even smaller than for a diode laser. (B.1.1) must be contrasted with (5.3.5). For a single photon mode taking P as the density of photons, $A_{ul}\Phi = B_{ul} = B_{lu}$, where Φ is the interaction volume. It is left to the reader to show that for atoms satisfying this type of interaction the number of electrons at any energy level $\mathscr{E}$ is $N_{\mathscr{E}} = N \exp(-\mathscr{E}/kT)$ (neglecting any degeneracy or multiplicity of states within an atom) for equilibrium with a correct Bose–Einstein distribution of photons.

The ideal rate equations for the changes in N_u and N_l are

$$dN_u/dt = D_u - A_{ul}N_u - B_{ul}P(N_u - N_l) \qquad \text{(B.1.2)}$$

$$dN_l/dt = A_{ul}N_u + B_{lu}P(N_u - N_l) - N_lA_{lL} + D_l \qquad \text{(B.1.3)}$$

The useful drive to the laser is $D_u = A_{Uu}N_U$, the total spontaneous decay rate from the upper level $\mathscr{E}_U$. The 'drive' D_l is undesirable and is neglected here; it can arise by electrons filling this lower state from other undetermined levels or by collisions with other atoms in a gas.

The rate equation for the uppermost level is

$$dN_U/dt = B_{LU}P_{pump}(N_L - N_U) - A_{Uu}N_U \qquad \text{(B.1.4)}$$

so, even with a high density of pump photons, N_U cannot exceed N_L, and the maximum level of $D_u = A_{Uu}N_L$. It is difficult then to suddenly change the pump rate or drive in such a four level laser as compared to an injection laser. In general N_L will take a value close to the equilibrium value because there will be many different upper levels able to populate this lowest level.

Lasing requires the net stimulated emission to overcome the photon losses and with $d/dt = 0$, from (B.1.2) and (B.1.3) it may be shown that

$$D_u = A_{lL} N_l \tag{B.1.5}$$

$$(N_u - N_l)(B_{ul} P + A_{ul}) = D_u[(A_{lL} - A_{ul})/A_{lL}] \tag{B.1.6}$$

For a good laser action the depopulation of the lower level should be at a high rate so that $A_{lL} \gg A_{ul}$.

Neglecting spontaneous emission for the output equation (B.1.1) it is seen that for $d/dt = 0$ and non-zero P,

$$N_u - N_l = 1/B_{ul}\tau_p = N_t \tag{B.1.7}$$

gives the 'threshold' condition corresponding to the threshold electron density n_t in the injection laser. Hence one finds approximately that the light output is given from

$$P/\tau_p = (D_u - A_{ul} N_t) \tag{B.1.8}$$

The drive has to reach a critical level to give lasing, and thereafter the drive electrons are turned into 'lasing' photons.

NOTES ON SOLUTIONS

1.1 $F_e = 40$; $S_e = 10$; $x = 4 \ln 10$ miles.

1.3 In 1.4.6 initial equilibrium values suggest $A = B = C/1.5$. Final equilibrium values would change to $B = 1.25A$, giving $P_{equ} = 4$, i.e. two more partners.

1.4 Note that $\sum\{nP(n, T)\} = \langle n \rangle$ and $\sum\{nP(n, T) - (n-1)P(n-1, T)\} = 0$. Hence $\mathrm{d}\langle n \rangle / \mathrm{d}T = p$. Similarly

$$\sum\{n(n-1)P(n, T) - (n-1)(n-2)P(n-1, T)\} = 0$$

and so

$$\mathrm{d}\langle n(n-1)\rangle / \mathrm{d}T = 2p \sum\{(n-1)P(n-1, T)\} = 2p\langle n \rangle$$

or $\mathrm{d}^2\langle n(n-1)\rangle/\mathrm{d}T^2 = 2p^2$. Integrating $\langle n \rangle = pT$, $\langle n(n-1)\rangle = (pT)^2$ $\langle (n - \langle n \rangle)^2 \rangle = \langle n^2 \rangle - \langle n \rangle^2$, $\langle n(n-1)\rangle = \langle n^2 \rangle - \langle n \rangle$; hence result.

1.5 If $I(t)$ has a step change then Q decays to its final value as $\exp(-bt)$, typically within 90% of its final value after time $2.3/b$. If $I(t)$ varies slowly on a time scale compared to this time then $Q = I/b$.

1.6 The *forced* variations of Q_1 or of Q_2 will be at the same rate as variations of the forcing drive $I(t)$. Hence, in second equation with k_4 large enough, $Q_2 = k_3 Q_1 / k_4$ eliminating Q_2.

2.1 Self collisions amongst electrons may change relative numbers with different energies although overall energy balance not affected. If τ_m is energy dependent then mobility can be changed by self collisions even though overall momentum balance is not changed.

2.2 Initial momentum = final momentum; initial energy − final energy = $\mathscr{E}_g$. Hence initial energy required is $\frac{1}{2}m^* v^2 = 3\mathscr{E}_g/2$.

2.3 $\sigma \approx 0.09(\mathrm{nm})^2$.

2.4 Phonon density N_p, $v_{\mathrm{electron}} \propto T^{1/2}$, $1/\tau_m \propto N_p \sigma_{pe} v_{\mathrm{electron}}$, so $\mu \propto T^{-3/2}$ if $N_p \sigma_{pe} \propto T$. (Note v_{electron} is thermal velocity.)

2.7 With n_d donor density $n_{tr} < n_d$, $n_{el} \ll n_d$; hence, from (2.3.12), with $\mathrm{d}/\mathrm{d}t = 0$, $n \ll (n_d)^2/N_{tr}$.

2.8 Mean increase in momentum $= \int_0^\infty FT \exp(-T/\tau)(\mathrm{d}T/\tau) = F\tau$: $F = eE$. Initial velocity $= 0$, so average distance travelled $= \int_0^\infty (F/2m^*)T^2) \exp(-T/\tau)(\mathrm{d}T/\tau) = (F/m^*)\tau^2$. Mean free time is τ; mean velocity is $eE\tau/m^*$.

3.1 $V_L^2 C_L R_s/(R_i + R_s)$ = excess energy lost per switch (on or off). 2 switches per 1 bit, 0 switches per 0 bit: average rate of energy loss = power $\sim B\mathscr{E}_{\mathrm{excess}}$.

3.2 Approximately 0.3 ns.

3.3 $Q = Q_0[1 - \chi \exp(-\chi) - \exp(-\chi)]$, where $\chi = t/(0.29RC)$; $\sim$13% improvement.

3.4 $\omega\tau = 1$ when $f \sim 250$ GHz.

3.5 $Q_0 = 2nC$ taking recombination lifetime $\tau_r = 200$ ns.

$$Q = 2\exp(-t/\tau_r) - 40[1 - \exp(-t/\tau_r)] - (200/495)[\exp(-t/T) - \exp(-t/\tau_r)]nC.$$

$Q = 0$ when $t = T_s \sim 11.5$ ns, $\tau_{\mathrm{snap}} \sim 57$ ps, after $Q = 0$.

3.6 (*a*) $R_s = 10\,\Omega$: mobility reduction by surface proximity, impurities in material to make it high resistivity, excessive photon energy compared to band gap making charge carriers 'hot' can change μ (sometimes even increase μ).

(*b*) $Q/\tau_r = q\eta R_{ph}$ gives net electronic charge by balancing rates of excitation and rates of loss. $R_s = L^2/[qR_{ph}\eta\tau_r(\mu_n + \mu_p)]$; $I = V/R_s$; hence G. Provided dark current is low enough then G can be forced larger than unity by enough voltage.

3.7 $Q = \rho_0 W/2$; $J = D(\rho_0/W)$; $\tau_b = Q/J = W^2/2D$; $\tau_b \sim 18$ ps.

4.2 $I_1 = 0$; $I_2 = Fn$.

4.4 Growth of 20 in charge approximately requires $3\tau_{nd}$. Roughly triangular field profile with peak field in 'centre' where $n = n_o$.

4.5 Voltage: peak square wave $\propto (K - 1)$; mean $\propto (1 + K)/2$. Current: peak square wave $\propto (1 - k)$; mean $(1 + k)/2$. Fundamental Fourier cosine component of unit square wave has peak amplitude $2/\pi$. Hence $\eta = (8/\pi^2)(K - 1)(1 - k)/(K + 1)(1 + k)$.

4.6 $\rho_1(t) \propto \exp(\alpha t - j\beta v_o t)$; substitute to find for small signals that $\alpha = (F\rho_0\mu_n/\varepsilon) - D\beta^2$. Diffusion slows down rate of growth. F

gives reduction of space charge fields in non-one-dimensional geometry (section 2.2.3). For $\beta \sim 2\pi/L$, taking L as the gate length of 1 μm and other reasonable values for FET channels, $\alpha \sim (F10^{12} - 10^{11})s < 0$ for $F < \frac{1}{10}$, easily obtained with metal gate close to surface channel. Any disturbances of charge must then decay with time.

5.1 48 μm, 6.2 THz; 187 μm, 1.6 THz; 3.6 mm, 83 GHz.
Gives simple boundaries where quantum statistics start to be important.

5.2
$$\mathrm{d}P_{12}/\mathrm{d}t + (P_{12} - P_{12e})/\tau_{12} = A(P_{12}+1)(P_2+1)P_1 - AP_{12}P_2(P_1+1)$$
$$\mathrm{d}P_1/\mathrm{d}t + (P_1 - P_{1e})/\tau_1 = A(P_1+1)P_2P_{12} - A(P_{12}+1)(P_2+1)P_1 = -\mathrm{d}P_2/\mathrm{d}t - (P_2 - P_{2e})/\tau_2$$

For equilibrium Bose-Einstein statistics apply provided energy conserved: $\hbar\omega_{12} = \hbar\omega_1 - \hbar\omega_2$. Writing $L_i = P/\tau_i$ as appropriate, equilibrium requires $L_{12\mathrm{out}} = L_{2\mathrm{out}} = L_{1\mathrm{in}}$, i.e. the photon count rate is the same. But $L = \mathrm{power}/\hbar\omega$; hence, the 'Manley-Rowe' relationships follow.

5.3 The probability is given from finding coefficient of F^P in the product $[1 - F/N]^N \to \exp F$. Coefficient of F^P in $\exp F$ is $F^P/P!$, which is the Poisson distribution for the probability of detecting P photons - apart from normalising factor which must be adjusted correctly.

5.4 Direct substitution in (5.3.5) gives required result. If $\mu > hf$, the condition for lasing, then the power has to grow. Energy would build up until some rate of escape matched rate of generation.

5.6 Poisson: two errors expected; chaotic: 47.6 million errors expected.

6.1 $\tau(\mathrm{d}I/\mathrm{d}t) + I = \eta q L$ (L in photons/s); $CR(\mathrm{d}V/\mathrm{d}t) + V = IR$. Both time constants 25 ps, half power frequency: $[1 + (\omega\tau)^2]^2 = 2$; $f \sim 4$ GHz.

6.2 Approximately 0.2 ps.

6.3 $CL = (1/GP_0)\tau_p$; $R/L = GP_0 + (1/\tau_r)$.

6.4 Unity.

6.5 $\tau_{rc} \ll \tau_r$ so for a step up in light input $I \propto 1 - \frac{65}{90}\exp(-t/\tau) - \frac{25}{90}\exp(-t/\tau_{rc})$.

Step response has a slow tail dependent on the rate for carriers to diffuse out of the contacts at a velocity of the order of $2D/W$, where W is thickness of the contact (see also problem 3.7).

7.1 $L = \{\Phi[\beta n_t(n_t - n_o)]^{1/2}/\tau_r\}$ photon/s when $I = q\Phi n_t/\tau_r$, where $n_t = n_0 + 1/G\tau_p$. At large L: $qL \approx I - I_t$, where $I_t = qn_t\Phi/\tau_r$.

7.2 $\Gamma = [(1/Q_L) + (1/Q_g) - (1/Q_R) - 2j\delta\omega]/[(1/Q_L) + (1/Q_R) - (1/Q_g) + 2j\delta\omega]$, where $Q = RC$ with appropriate subscripts. Oscillation if Γ infinite, i.e. reflects finite power even though none is going in! Oscillation if $(1/Q_L) + (1/Q_R) = (1/Q_g)$ or $G = (1/\tau_L) + (1/\tau_R)$.

7.4 $k = (1/\tau_L)^{1/2}$ to give amplitude growth $\exp(t/2\tau_L)$ as power is fed back in under time reversal.

REFERENCES

There are just too many books and papers touching the breadth of subjects covered here so no attempt has been made to make a comprehensive reference list. In general the reader who wishes to follow up a subject needs a further introduction. Books rather than papers tend to give such introductions to topics - even if a book is initially at too advanced a level for the reader. A selection of references has been made here and no intentional slight should be taken for the omission of a reader's favourite text. As background, the reader must understand differential equations, and almost any good applied mathematics book will cover the required mathematics (see, for example, reference 6). For background level in device physics, then many of the basic topics in reference 5, or, of course, 9, will provide a more than adequate starting point. In electromagnetic theory, reference 12 is helpful for the initial chapters, but a more advanced background (reference 75) is required for chapter 7. Background material for electronic circuits is well covered by references 16 or 32.

1 Kompfner, R., *in* Lindsay, P. A. *Introduction to quantum mechanics for electrical engineers*, McGraw-Hill, 1967

2 *Nuffield advanced science chemistry*, (students book II (topic 14)), Longman, 1984

3 French, A. P., and Taylor, E. F., *Introduction to quantum physics*, Norton, 1978

4 Dirac, P. A. M., *Principles of quantum mechanics*, 4th edn, Clarendon, Oxford, 1968

5 Solymar, L., and Walsh, D., *Lectures on properties of materials*, 3rd edn, Oxford University Press, 1984

6 Kreysig, E., *Advanced engineering mathematics*, 5th edn, Wiley, 1983

7 Hamming, R. W., *Coding and information theory*, Prentice-Hall, 1980

8 Sze, S. M., *Physics of semiconductor devices*, 2nd edn, Wiley Interscience, 1981

9 Carroll, J. E., *Physical models of semiconductor devices*, Arnold, 1974

10 Moll, J. L., *Physics of semiconductors*, McGraw-Hill, 1964

11 Carroll, J. E., *Hot electron microwave generators*, Arnold, 1970

12 Oatley, C. W., *Electric and magnetic fields*, Cambridge University Press, 1976

13 Beck, A. H. W., *Space charge waves*, Pergamon, 1958, pp. 111–12

14 Pankove, J. I., *Optical processes in semiconductors*, Prentice-Hall, 1971, (corrected Dover reprint 1975)
15 Hobson, G. S., *Charge transfer devices*, Arnold, 1978
16 Smith, R. J., *Circuits, devices and systems*, 4th edn, Wiley, 1984
17 Steidel, C. A., *in* Sze, S. M. (Ed.) *VLSI technology*, McGraw-Hill, 1983, chap. 13
18 Carroll, J. E., *Fast switching in semiconductors, Science Progress*, 1984, **69**, pp. 101–27
19 Olson, H. M., Barber, M. R., Sodomsky, K. F., and Zacharias, A., p–i–n diode/microwave switches *in* Watson, H. A. *Microwave semiconductor devices and their circuit applications*, McGraw-Hill, 1969, Chap. 9/10
Irvin, J. C., Lee, T. P., Decker, D. R., Uenohara, M., and Gewartowski, J. W., Varactors and applications *in* Watson, H. A., *op. cit.*, Chap. 7/8
Miller, L. E., and Engelbrecht, R. S., Microwave transistors *in* Watson, H. A., *op. cit.*, Chap. 17/18
20 Gentry, F. E., Gutzweiler, F. W., Holynyak, N., and Von Zastrow, E. E., *Semiconductor controlled rectifiers*, Prentice-Hall, 1964
21 Auston, D. H., Picosecond non linear optics, *in* Shapiro, S. L., *Ultra-short light pulses*, Springer Verlag, 1977, chap. 4
See also articles by Auston and others in Hochstrasser, R. M., Kaiser, W., and Shank, C. V. (Eds.) *Picosecond phenomena II*, Pt. II *Advances in optoelectronics*, Springer Verlag, 1980 and *IEEE J. Quantum Electron.*, 1983, **QE-19**, (4)
22 Gibbons, G., *Avalanche diode microwave oscillators*, Oxford University Press, 1973
23 Gowar, J., *Optical communication systems*, Prentice-Hall, 1984
24 Rhoderick, E. H., *Metal-semiconductor contacts*, Clarendon Press, Oxford, 1977
25 Wang, S. Y., and Bloom, D. M., 100 GHz bandwidth planar GaAs Schottky photodiode, *Electron. Lett.*, 1983, **19**, pp. 554–5
26 Pengelly, R. S., *Microwave field effect transistors theory, design and applications*, Research Studies Press, 1982
27 Shannon, J. M., Hot electron diodes and transistors, *in* Rhoderick, E. H. (Ed.) *Solid state devices*, Institute of Physics conference series no. 69, 1983, pp. 45–62
28 Shockley, W., Unipolar field effect transistors, *Proc. I.R.E.*, 1952, **40**, p. 1365
29 Gossard, A. C., Quantum well structures and superlattices *in* Rhoderick, E. H. (Ed.) *Solid state devices*, Institute of Physics conference series no. 69, 1983, pp. 1–14
30 Beaufoy, R., and Sparkes, J. J., The junction transfer as a charge controlled device, *ATE J. (London)*, 1957, **13**, pp. 310–24
31 Gummel, H. K., and Poon, H. C., An integral charge control model of bipolar transistors, *Bell Syst. Tech. J.*, 1970, **49**, pp. 827–51
32 Ahmed, H., and Spreadbury, P. J., *Digital and analogue electronics for engineers*, 2nd edn, Cambridge University Press, 1984
33 Lindmayer, J., and Wrigley, C. Y., 'Fundamentals of semiconductor devices', Van Nostrand, 1965, Chap. 5
34 Bozler, C. O., and Alley, G. D., Fabrication and numerical simulation of the permeable base transistor, *IEEE Trans.*, 1980, **ED-27**, pp. 1128–44
35 Welbourn, A. D., *Gigabit logic: a review, IEE Proc. I, Solid-State, Electron Dev.*, 1982, **129**, pp. 157–72
36 Duncan, T., *Physics*, Murray, 1982, section 19
37 Jeans, J., *An introduction to the kinetic theory of gases*, Cambridge University Press, 1940

38 Huang, K., *Statistical mechanics*, Wiley, 1963
39 Beck, A. H. W., *Statistical mechanics, fluctuations and noise*, Arnold, 1976
40 Landau, L. D., and Lifshitz, E. M., *Statistical physics*, 2nd edn, Pergamon, 1969
41 Goldstein, H., *Classical mechanics*, 2nd edn, Addison-Wesley, 1980
42 Gunn, J. B., Microwave oscillations of current in III–V semiconductors, *Solid State Commun.*, 1963, **1**, p. 88
43 Hilsum, C., Transferred electron amplifiers and oscillators, *Proc. Inst. Rad. Engrs.*, 1962, **50**, p. 185
44 Ridley, and B. K., Watkins, T. B., The possibility of negative resistance effects in semiconductors, *Proc. Phys. Soc.* (*London*), 1971, **78**, p. 293
45 Constant, E., *Modelling of sub-micron devices*, *in* Carroll, J. E. (Ed.) *Institute of Physics conference series no.* 57, 1980
46 Liechti, C. A., Microwave field effect transistors, *IEEE Trans.*, 1976, **MTT-24**, p. 279
47 Jacoboni, C., and Reggiani, L., The Monte Carlo method for the solution of charge transport in semiconductors with applications to covalent materials, *Rev. Mod. Phys.*, 1983, **55**, pp. 645–705
48 Bowler, M. G., *Lectures on statistical mechanics*, Pergamon, 1982
49 Landsberg, P. T., *The enigma of time*, Hilger, 1982
50 Huang, K., *Statistical mechanics*, Wiley, 1963
51 Kittel, C., and Kroemer, H., *Thermal physics*, 2nd edn, Freeman, 1980
52 Park, D., *The quantum theory*, McGraw-Hill, 1974
53 Marcuse, D., *Principles of quantum electronics*, Academic Press, 1980
54 Loudon, R., *The quantum theory of light*, 2nd edn, Oxford University Press, 198
55 Landsberg, P. T., Photons at non-zero chemical potential, *J. Phys. C, Solid State Phys.*, 1981, **14**, L1025–27
56 Wilson, J., and Hawkes, J. F. B., *Optoelectronics*: *an introduction*, Prentice-Hall, 1983.
(Provides introductions to a wide range of devices.)
57 Siegman, A. E., *Introduction to masers and lasers*, McGraw-Hill, 1971
58 Yariv, A., *Quantum electronics*, 2nd edn, Wiley, 1975
59 Sargent, M., Scully, M. O., and Lamb, W. E., *Laser physics*, Addison Wesley, 1974
60 Bergh, A. A., and Dean, P. J., *Light emitting diodes*, Oxford University Press, 1976
61 Pearsall, T., *GaInAsP alloy semiconductors*, Wiley, 1982
62 Casey, H. C., and Panish, M. B., *Heterostructure lasers*, Academic Press, 1978
63 Thompson, G. H. B., *Physics of semiconductor laser devices*, Wiley, 1980
64 Adams, M. J., *An introduction to optical waveguides*, Wiley, 1981
65 Adams, M. J., and Osinki, M., Longitudinal mode competition in semiconductor lasers – rate equations revisited, *IEE Proc. I, Solid-State and Electron Dev.*, 1982, **129** pp. 271–4
66 Epworth, R., Modal noise – causes and cures, *Laser Focus*, **17** (9), pp. 109–15
67 Renner, D., and Carroll, J. E., The effect of spontaneous emission coupling in semiconductor lasers, *Electron. Lett.*, 1978, **14**, pp. 779–80
68 Peterman, K., and Arnold, G., Noise and distortion characteristics of semiconductor lasers in optical fibre communication systems, *IEEE J. Quantum Electron.*, 1982, **QE-18**, pp. 543–55
69 Gowar, J., *Optical communication systems*, Prentice-Hall, 1984
70 Demokan, M. S., *Mode locking in solid state and semiconductor lasers*, Research Studies Press, 1982. (For an elementary introduction see also reference 56.)

71 Haus, H., Models of mode locking in a laser diode in an external resonator, *Proc. IEE, Pt. I*, 1980, **127**

72 Haus, H. A., Theory of modelocking of a laser diode in an external resonator, *J. Appl. Phys.*, 1980, **51**, pp. 4042–9

73 Fork, R. L., Shank, C. V., Yen, R., and Hirlimann, C. A., Femtosecond optical pulses, *IEEE J. Quantum Electron.*, 1983, **QE-19**, (4)

74 Sander, K. F., and Reed, G. A. L., *Transmission and propagation of electromagnetic waves*, Cambridge University Press, 1978, (see, for example, pp. 123, 343)

75 Ramo, S., Whinnery, J. R., and Van Duzer, T., *Fields and waves in communication electronics*, 2nd edn, Wiley, 1968

76 Slater, J. C., *Microwave electronics*, Van Nostrand, 1950 (Dover reprint 1969)

77 Marcuse, D., *Principles of quantum electronics*, Academic Press, 1980

78 Bracewell, R., *The Fourier transform and its applications*, 2nd edn, McGraw-Hill, 1978

79 Bimberg, D., Ketterer, K., Scholl, E., and Volmer, H. P., *Avalanche generator triggered picosecond light pulses from unbiased GaAs/GaAlAs lasers*, Proc. 14th ESSDERC Physica, 1985

80 Ippen, E., Eilenberger, D. J., and Dixon, R. W., Picosecond pulse generation by passive mode locking of diode lasers, *Appl. Phys. Lett.*, 1980, **37**, p. 267

81 Margenau, H., and Murphy, G. M., *The mathematics of physics and chemistry*, 2nd edn, Van Nostrand, 1956

82 Van der Ziel, J. P., Spectral broadening of pulsating $Al_xGa_{1-x}As$ double hetero-structure lasers, *IEEE J. Quantum Electron.*, 1979, **QE-15**, (11), pp. 1277–81

83 Henry, C., Theory of the linewidth of semiconductor lasers, *IEEE J. Quantum Electron.* 1982, **QE-18**, (2) pp. 259–64

INDEX